Soldier Training Publication
No. 21-1-SMCT

Washington, DC, 28 September 2017

Contents

Distribution Restriction: Approved for public release; distribution is unlimited.
*This manual supersedes STP 21-1-SMCT, dated 10 August 2015.
STP 21-1-SMCT   i

Preface

This manual is one of a series of soldier training publications (STPs) that support individual training. Commanders, trainers, and Soldiers will use this manual and STP 21-24-SMCT to plan, conduct, sustain, and evaluate individual training of warrior tasks and battle drills in units.

This manual includes the Army Warrior Training plan for warrior skills level (SL) 1 and task summaries for SL 1 critical common tasks that support unit wartime missions. This manual is the only authorized source for these common tasks. Task summaries in this manual supersede any common tasks appearing in military occupational specialty (MOS)-specific Soldier manuals.

Training support information, such as reference materials, is also included. Trainers and first-line supervisors will ensure that SL 1 Soldiers have access to this publication in their work areas, unit learning centers, and unit libraries.

This manual applies to the Active Army, the Army National Guard/Army National Guard of the United States, and the U.S. Army Reserve unless otherwise stated.

The proponent of this publication is the United States Army Training and Doctrine Command (TRADOC), with the United States Army Training Support Center (ATSC) designated as the principle publishing, printing, and distribution agency. Proponents for the specific tasks are the Army schools and agencies as identified by the school code, listed in appendix A. This code consists of the first three digits of the task identification number.

Record any comments or questions regarding the task summaries contained in this manual on a DA Form 2028 (*Recommended Changes to Publications and Blank Forms*) and send it to the respective task proponent, James Rose, CIMT, james.a.rose20.civ@mail.mil, with information copies forwarded to—

> • Commander, U.S. Army Training and Doctrine Command
> ATTN: ATCG-MT
> Fort Eustis, VA 23604-5701
>
> • Commander, U.S. Army Training Support Center
> ATTN: ATIC-APR
> Fort Eustis, VA 23604-5166

This page intentionally left blank.

CHAPTER 1

Introduction to the SMCT System

1-1. GENERAL

The Army's basic mission is to train and prepare Soldiers, leaders, and units to fight and win in combat. As explained in the Army's capstone training doctrine (ADP 7-0), units do not have the time or the resources to achieve and sustain proficiency with every possible training task. Therefore, commanders must identify the tasks that are the units' critical wartime tasks. These tasks then become the unit's Mission Essential Task List (METL). Commanders use the METL to develop their unit-training plan. Noncommissioned officers (NCOs) plan the individual training that Soldiers need to become warriors and to accomplish the METL. The STPs, also known as Soldier's manuals (SMs), provide the critical individual tasks for each military occupational specialty (MOS) that support all of the unit's missions. The NCO leadership uses the tasks in the SMs to train the Soldiers and measure the Soldiers' proficiency with these unit-critical tasks. The manuals provide task performance and evaluation criteria and are the basis for individual training and evaluation in the unit and for task-based evaluation during resident training.

The Army identified warrior tasks and battle drills (WTBD) that enhance a Soldier's readiness to fight on the battlefield. Warrior tasks are a collection of individual Soldier skills known to be critical to Soldier survival. Examples include weapons training, tactical communications, urban operations, and first aid. Battle drills are group skills designed to teach a unit to react and survive in common combat situations. Examples included react to ambush, react to chemical attack, and evacuate injured personnel. WTBD increases the relevance of training to current combat requirements and enhance the rigor in training. The driving force behind the change comes from lessons learned. Standards remain constant but commanders must be aware that the enemy adapts at once and Soldier training will change sooner because of current operational environments.

Note: If a task identified in the SMCT is not current refer to "DTMS", or the Central Army Registry (CAR) https://atiam.train.army.mil/catalog/catalog/search.html, all tasks are reviewed every two years and may change before the SMCT is updated.

1-2. PURPOSE

This Soldier's Manual of Common Tasks (SMCT), Warrior Skill Level (SL) 1, contains the individual tasks that are essential to the Army's ability to win on the modern battlefield. In an operational environment, regardless of job or individual MOS, each Soldier risks exposure to hostile actions. This manual contains the warrior skills that Soldiers must be able to perform to fight, survive, and win in combat.

This SMCT gives the commander, NCO trainer, first-line supervisor, and individual Soldiers the information necessary to support integration and sustainment training in their units. This information allows trainers to plan, prepare, train, evaluate, and monitor individual training of warrior tasks. Using the appropriate mission-training plan (MTP), military occupational specialty (MOS)-specific Soldier's training publication (STP), and this manual helps provide the foundation for an effective unit-training plan.

1-3. COMMANDER'S RESPONSIBILITIES

The commander at each level develops a unit METL in consultation with the command sergeant major and subordinate commanders. Using the training planning process described in ADP 7-0, the commander develops the METL and then determines the level of training needed to attain and maintain proficiency. WTBD in Chapter 3 supports an Army at war and becomes the key element in Army Warrior Training (AWT). Commanders use the unit METL and AWT to determine the necessary training for the unit and develop a strategy to accomplish the required training throughout the fiscal year (FY). The commander also gives the NCO leadership the guidance they need to carry out this strategy. Each commander must design a unit training plan that prepares the unit for the full spectrum of operations. Soldiers must develop and sustain proficiency in the critical tasks for their MOS and skill level. The commander's unit training program should provide individual training for all Soldiers assigned to the unit and evaluate Soldier proficiency by routine. The leader's assessment and the AWT are two tools that give the NCO leadership and commander information about the status of training for individuals and for the unit, which should be integrated with collective training such as the MTPs, crew drills, and battle drills.

Chapter 2 provides information about where tasks are trained to standard and how often tasks are trained to maintain proficiency.

Based on the commander's guidance, individual training in the unit is the responsibility of the NCO trainers. The commander must give the NCO trainer the priorities, resources, and directions needed to carry out training. He or she must also assess the training results of the MTP and other training events, and adjust the unit training plan as a result. To develop a training program, use the following seven-step approach:

Step 1. Set the objectives for training.

Step 2. Plan the resources (personnel, time, funds, facilities, devices, and training aids).

Step 3. Train the trainers.

Step 4. Provide the resources.

Step 5. Manage risks, environmental and safety concerns.

Step 6. Conduct the training.

Step 7. Evaluate the results.

1-4. TRAINER'S RESPONSIBILITIES

Trainers must use the following steps to plan and evaluate training:

a. *Identify individual training requirements.* The NCO determines which tasks Soldiers need to train based on the commander's training strategy. The unit's training plan, METL, MTP, and the AWT plan (Chapter 2) are sources for helping the trainer define the individual training needed.

b. *Plan the training.* Plan individual training based on the unit's training plan. Be prepared to take advantage of opportunities to conduct individual training ("hip pocket" training).

c. *Gather the training references and materials.* The task summaries list references that can assist the trainer in preparing for the training of that task. The Reimer Digital Library provides current training materials.

d. *Manage risks and environmental and safety concerns.* Assess the risks involved with training a specific task regarding the conditions current at the time of training and, if necessary, implement controls to reduce the risk level. Ensure that training preparation takes into account those cautions, warnings, and dangers associated with each task as well as environmental and safety concerns (ATP 5-19).

e. *Train each Soldier.* Demonstrate to the Soldier how to do the task with standard proficiency and explain (step by step) how to do the task. Give each Soldier the opportunity to practice the task step by step.

f. *Check each Soldier.* Evaluate how well each Soldier performs the tasks in this manual. Conduct these evaluations during individual training sessions or while evaluating individual proficiencies when conducting unit collective tasks. This manual provides a training and evaluation guide for each task to enhance the NCO's ability to conduct year-round, hands-on evaluations of tasks critical to the unit's mission. Use the information in the AWT plan (chapter 2) as a guide to determine how often to train Soldiers using each task to maintain proficiency.

g. *Record the results.* Use the leader book referred to in ADP 7-0 to record task performance. This gives the leader total flexibility with the methods of recording training tasks. The trainer may use DA Form 5164-R (*Hands-on Evaluation*) and DA Form 5165-R (*Field Expedient Squad Book*) as part of the leader book. These forms are optional and reproducible anywhere.

h. *Retrain and evaluate.* Work with each Soldier until he/she performs the task to standard. Well-planned, integrated training increases the professional competence of each Soldier and contributes to the development of an efficient unit. The NCO or first-line supervisor is a vital link to the conduct of training.

1-5. SOLDIER'S RESPONSIBILITIES

Each Soldier must be able to perform the individual tasks that the first-line supervisor has identified based on the unit's METL. The Soldier must perform the task to the standard listed in this SMCT. If a Soldier has a question about how to do a task, or which tasks in this manual he or she must perform, it is the Soldier's responsibility to go to the first-line supervisor for clarification. The first-line supervisor knows how to perform each task or can direct the Soldier to the appropriate training materials. In addition, each Soldier should—

    a. Know the training steps for both the WTBD and the MOS-specific critical tasks for his or her skill level. A list of the critical tasks is found in chapter 2 of this manual and the STP for the specific MOS (MOS-specific tasks). Check DTMS or the Central Army Registry (CAR) for new training materials to support self-development with maintaining earlier trained tasks or to learn new tasks.

1-6. TASK SUMMARIES

Task summaries document the performance requirements of a critical warrior task. They provide the Soldier and the trainer with the information necessary to evaluate critical tasks. The formats for the task summaries are—

    a. *Task title.* The task title identifies the action to perform.

    b. *Task number.* The task number is an l0-digit number or letters that identifies each task. The first three digits of the number represent the proponent code for that task. (Appendix A provides a list of proponent codes.) Include the entire 10-digit task number, along with the task title, in any correspondence relating to the task.

    c. *Conditions.* The task conditions identify all the equipment, tools, materials, references, job aids, and supporting personnel that the Soldier needs to perform the task. This section identifies any environmental conditions that can alter task performance such as visibility, temperature, or wind. This section also identifies any specific cues or events (for example, a chemical attack or identification of an unexploded ordnance hazard) that trigger task performance.

    d. *Standards.* A task standard specifies the requirements for task performance by indicating how well, complete, or accurate a product must be produced, a process must be performed, or both. Standards are described in terms of accuracy, tolerances, completeness, format, and clarity, number of errors, quantity, sequence, or speed of performance.

    e. *Training and evaluation guide.* This section has two parts. The first part, Performance Steps, lists the individual steps that the Soldier must complete to perform the task. The second part is the Performance Evaluation Guide. This provides guidance about how to evaluate a Soldier's performance of the task. It is composed of three subsections. The *Evaluation Preparation* subsection identifies special setup procedures and, if required, instructions for evaluating the task performance. Sometimes the conditions and standards must be modified so that the task can be evaluated in a situation that does not, without approximation, duplicate actual field performance. The *Performance Measures* subsection identifies the criteria for acceptable task performance. The Soldier is rated (GO/NO- GO) on how well he or she performs specific actions or produces specific products. As indicated in *Evaluation Guidance*, a Soldier must score a GO on all or specified performance measures to receive a GO on the task in order to be considered trained.

f. *References.* This section identifies references that provide more detailed and thorough explanations of task performance requirements than that given in the task summary description. This section identifies resources the Soldier can use to improve or maintain performance.

g. In addition, task summaries can include safety statements, environmental considerations, and notes. Safety statements (danger, warning, and caution) alert users to the possibility of immediate death, personal injury, or damage to equipment. Notes provide additional information to support task performance.

1-7. TRAINING TIPS FOR NCO LEADERS

a. Prepare yourself.

(1) Get training guidance from your chain of command about when to train, which Soldiers to train, availability of resources, and a training site.

(2) Get task, conditions and standards from the task summary in this manual. Ensure that you can do the task. Review the task summary and the references in the reference section. Practice doing the task or, if necessary, have someone train you how to perform the task.

b. Prepare the resources.

(1) Obtain the required resources as identified in the conditions statement for each task and/or modified in the training and evaluation guide.

(2) Gather the equipment and ensure that it is operational.

(3) Prepare a training outline consisting of informal notes about what you want to cover during your training session.

(4) Practice your training presentation.

(5) Coordinate for the use of training aids and devices.

(6) Prepare the training site using the conditions statement as modified in the training and evaluation guide.

c. Train the Soldiers.

(1) Tell the Soldier what task to do and how well it must be done. Refer to the task standards and the performance measures for the task, as appropriate.

(2) Caution Soldiers about safety, environment, and security considerations.

(3) Demonstrate how to do the task to the standard level. Have the Soldiers study the appropriate training materials.

(4) Provide any necessary training involving basic skills Soldiers must have before they can be proficient with the task.

(5) Have the Soldiers practice the task until they can perform it to standard levels.

(6) Provide critical information to those Soldiers who fail to perform at task standard levels, and have them continue to practice until they can perform at standard levels.

(7) Combine training involving the individual tasks contained in this manual with the collective tasks contained in the MTP. Ensure that the necessary safety equipment and clothing needed for proper performance of the job are on hand at the training site.

d. Record the results: First-line supervisors record the results and report information to the unit leadership.

1-8. TRAINING SUPPORT

Appendix A lists the task proponents and agency codes (first three digits of the task number) with addresses for submitting comments concerning specific tasks in this manual.

1-9. EVALUATING TASK PERFORMANCE

Trainers need to keep the following points in mind when preparing to evaluate their Soldiers:

a. Review the performance measures to become familiar with the criteria about which you will score the Soldier.

b. Ensure that all necessary equipment and clothing needed for proper performance of the job are on hand at the training site. Remember to include safety equipment.

c. Prepare the test site according to the conditions section of the task summary. Some tasks contain special evaluation preparation instructions. These instructions tell the trainer what modifications must be made concerning job conditions to evaluate the task. Reset the site to its original condition after evaluating each Soldier to ensure that the conditions are the same for each Soldier.

d. Advise each Soldier about any special guidance that appears in the evaluation preparation section of the task summary before evaluating.

e. Score each Soldier regarding the information in the performance measures and evaluation guidance. Record the date of training and task performance score (GO / NO GO) in the sections training records for each Soldier.

(1) When applicable, conduct an exercise after-action review to allow training participants to discover for themselves what happened, why it happened, and how it can be done better. Once all key points are discussed and linked to future training, the evaluator will make the appropriate notes for inclusion into the score.

(2) Score the Soldier GO if all performance measures pass. Score the Soldier NO GO if the Soldier fails any step. If the Soldier fails, Show the Soldier what they did wrong and allow the Soldiers to take the test again.

This page intentionally left blank.

CHAPTER 2

Training Guide

2-1. THE ARMY WARRIOR TRAINING PLAN

a.   Army Warrior Training focuses on training Soldiers warrior tasks, battle drills, and tasks from a unit's METL. This chapter and chapter 3 provides information identifying individual tasks to train and assist in the trainer's planning, preparation, training assessment, and monitoring of individual training in units. It lists by general subject area, and skill level, the critical warrior tasks Soldiers must perform, the initial training location, and a suggested expertise of training.

b.   The training location column uses brevity codes to indicate where the task is first taught to standard levels. If the task is taught in the unit, the word "UNIT" appears in this column. If the task is trained by a self-development media, "SD" appears in this column. If the task is taught in the training base, the brevity code (BCT, OSUT, and AIT) of the resident course appears. Brevity codes and resident courses are listed below.

Brevity Codes	
BCT	Basic Combat Training
OSUT	One Station Unit Training
AIT	Advanced Individual Training
UNIT	Trained in/by the Unit
SD	Self-Development Training

c.   The sustainment-training column lists how often (frequency) Soldiers should train with the task to ensure they maintain their proficiency. This information is a guide for commanders to develop a comprehensive unit training plan. The commander, in conjunction with the unit trainers, is in the best position to determine which tasks, and how often Soldiers should train to maintain unit readiness. (See chapter 3 for a list of individual tasks that support the WTBD to be trained in each Army unit.)

Frequency Codes	
AN	Annually
SA	Semiannually
QT	Quarterly

Army Warrior Training Plan			
Task Number	*Title*	*Training Location*	*Sustainment Training Frequency*
Warrior Skill Level 1			
Subject Area 1: Shoot/Maintain, Employ, and Engage Targets with Individually Assigned Weapon System:			
071-COM-0032	Maintain an M16 Series Rifle/M4 Series Rifle Carbine	BCT/OSUT	AN
071-COM-0029	Perform a Function Check on an M16-Series Rifle/M4 Series Carbine	BCT/OSUT	QT
071-COM-0028	Load an M16- Series/M4 Series Carbine	BCT/OSUT	SA
071-COM-0027	Unload an M16- Series Rifle/M4Series Carbine	BCT/OSUT	SA
071-COM-0033	Correct Malfunctions of an M16-Series Rifle/M4 Series Carbine	BCT/OSUT	SA
071-COM-0031	Zero an M16-Series Rifle/M4 Series Carbine	BCT/OSUT	SA
071-COM-0030	Engage Targets with an M16-Series Rifle/ M4 Series Carbine	BCT/OSUT	SA
Subject Area 2: Shoot/Employ Hand Grenades:			
071-COM-4401	Perform Safety Checks on Hand Grenades	BCT/OSUT	AN
071-COM-4407	Employ Hand Grenades	BCT/OSUT	AN
Subject Area 3: Move/ Perform Individual Movement Techniques:			
071-COM-0541	Perform Exterior Movement Techniques during an urban operation	BCT/OSUT	AN
071-COM-0503	Move Over, Through, or Around Obstacles (Except Minefields)	BCT/OSUT	SA
071-COM-1000	Identify Topographic Symbols on a Military Map	BCT/OSUT	AN
071-COM-1001	Identify Terrain Features on a Map	BCT/OSUT	AN
071-COM-1002	Determine the Grid Coordinates of a Point on a Military Map	BCT/OSUT	AN

Army Warrior Training Plan			
Task Number	**Title**	**Training Location**	**Sustainment Training Frequency**
071-COM-1008	Measure distance on a Map	BCT/OSUT	AN
071-COM-1005	Determine a Location on the Ground by Terrain Association	BCT/OSUT	AN
071-COM-1012	Orient a Map to the Ground by Map-Terrain Association	BCT/OSUT	AN
071-COM-1011	Orient a Map Using a Lensatic Compass	BCT/OSUT	AN
071-COM-1003	Determine a Magnetic Azimuth Using a Lensatic Compass	BCT/OSUT	AN
071-COM-0018	Determine an Azimuth Using a Protractor	BCT/OSUT	AN
071-COM-1016	Convert an Azimuth	BCT/OSUT	AN
071-COM-0017	Compute Back Azimuth	Unit	AN
071-COM-1006	Navigate from One Point on the Ground to Another Point While Dismounted	BCT/OSUT	AN
071-COM-1014	Locate an Unknown Point on the Ground by Intersection	BCT/OSUT	AN
071-COM-1015	Locate an Unknown Point on the Ground by Resection	BCT/OSUT	SA
071-COM-0501	Move as a Member of a Team	BCT/OSUT	SA
071-COM-0502	Move Under Direct Fire	BCT/OSUT	SA
071-COM-0510	React to Indirect Fire While Dismounted	BCT/OSUT	SA
071-COM-0513	Select Hasty Fighting Positions	BCT/OSUT	SA
Subject Area 4: Communicate:			
113-COM-2070	Operate Single Channel Ground and Airborne Radio System (SINCGARS) Single-Channel (SC)	BCT/OSUT	SA
113-COM-1022	Perform Voice Communications	BCT/OSUT	AN
081-COM-0101	Request Medical Evacuation	BCT/OSUT	SA

Army Warrior Training Plan			
Task Number	**Title**	**Training Location**	**Sustainment Training Frequency**
171-COM-4079	Send a Situation Report (SITREP)	BCT/OSUT	AN
171-COM-4080	Send a Spot Report (SPOTREP)	BCT/OSUT	AN
071-COM-0608	Use Visual Signaling Techniques	BCT/OSUT	SA
Subject Area 5: Survive:			
031-COM-1010	Maintain Your Assigned Protective Mask	BCT/OSUT	AN
031-COM-1004	Protect Yourself from Chemical and Biological (CB) Contamination Using Your Assigned Protective Mask	BCT/OSUT	AN
031-COM-1007	React to Chemical or Biological (CB) Hazard/Attack	BCT/OSUT	AN
031-COM-1005	Protect Yourself from CBRN Injury/Contamination by Assuming MOPP Level 4	BCT/OSUT	SA
031-COM-1009	Detect Chemical Agents Using M9 Detector paper	BCT/OSUT	AN
031-COM-1008	Identify Liquid Chemical Agents Using M8 paper	BCT/OSUT	AN
031-COM-1006	Decontaminate Yourself and Individual Equipment Using Chemical Decontaminating Kits	BCT/OSUT	AN
031-COM-1001	React to Nuclear Hazard / Attack	BCT/OSUT	AN
031-COM-1003	Mark CBRN-Contaminated Areas	Unit	AN
081-COM-1001	Evaluate a Casualty (Tactical Combat Casualty Care)	BCT/OSUT	AN
081-COM-1005	Perform First Aid to Prevent or Control Shock	BCT/OSUT	AN
081-COM-1023	Perform First Aid to Open the Airway	BCT/OSUT	AN
081-COM-1054	Apply an Emergency Bandage	BCT/OSUT	AN
081-COM-0099	Apply a Hemostatic Dressing	BCT/OSUT	AN

Army Warrior Training Plan			
Task Number	Title	Training Location	Sustainment Training Frequency
081-COM-0069	Apply a Occlusive Dressing	BCT/OSUT	AN
081-COM-0048	Apply a Combat Application Tourniquet (CAT)	BCT/OSUT	AN
081-COM-1055	Apply a FOX Eye Shield	BCT/OSUT	AN
081-COM-0013	Initiate a DD Form 1380 Tactical Combat Casualty Care (TCCC) Card	BCT/OSUT	AN
081-COM-1046	Transport a Casualty	BCT/OSUT	AN
081-COM-1007	Perform First Aid for Burns	BCT/OSUT	AN
081-COM-0101	Request Medical Evacuation	BCT/OSUT	AN
052-COM-1271	Identify Visual Indicators of an Improvised Explosive Device (IED)	BCT/OSUT	AN
052-COM-1270	React to Possible Improvised Device (IED) Attack	BCT/OSUT	AN
071-COM-0815	Practice , Noise, Light, and Litter Discipline	BCT/OSUT	AN
071-COM-0804	Perform Surveillance without the Aid of Electronic Device	Unit	SA
071-COM-0801	Challenge Persons Entering your Area	BCT/OSUT	SA
301-COM-1050	Report Information of Potential Intelligence Value	BCT/OSUT	SA
071-COM-1004	Perform Duty as a Guard	BCT/OSUT	SA
052-COM-1361	Camouflage Yourself and Individual Equipment	BCT/OSUT	AN
071-COM-4408	Construct an Individual Fighting Position	BCT/OSUT	AN
159-COM-2026	Identify Combatant and Non-Combatant Personnel & Hybrid Threats	BCT/OSUT	AN
181-COM-1001	Conduct operations According to the Law of War	BCT/OSUT	AN

Army Warrior Training Plan			
Task Number	*Title*	*Training Location*	*Sustainment Training Frequency*
171-COM-0011	Employ Progressive Levels of Individual Force	BCT/OSUT	AN
191-COM-0009	Search a Detainee	BCT/OSUT	AN
071-COM-0512	Perform Hand-to-Hand Combat	BCT/OSUT	AN
Battle Drill 1	Battle Drill: React to Contact	BCT/OSUT/ UNIT	AN
N/A	Engage Targets with your assigned weapon	BCT/OSUT	AN
071-COM-0513	Select Hasty Fighting Positions (Repeat)	BCT/OSUT	SA
071-COM-0608	Use Visual Signaling Techniques (Repeat)	BCT/OSUT	AN
071-COM-0502	Move under Direct Fire (Repeat)	BCT/OSUT	SA
071-COM-0510	React to Indirect Fire While Dismounted(Repeat)	BCT/OSUT	SA
113-COM-1022	Perform Voice Communications (Repeat)	BCT/OSUT	AN
071-COM-0501	Move as a member of a Team (Repeat)	BCT/OSUT	SA
071-COM-4407	Employ Hand Grenades (Repeat)	BCT/OSUT	AN
071-COM-0503	Move over, through, or around obstacles (except minefields)	BCT/OSUT	AN
052-COM-1271	Identify Visual Indicators of an Improvised Explosive Device (Repeat)	BCT/OSUT	AN
051-COM-1270	React to Possible Improvised Explosive Device (Repeat)	BCT/OSUT	AN
191-COM-5148	Search an Individual in a Tactical Environment	BCT/OSUT	AN
Battle Drill 2	Battle Drill: Establish Security at the Halt	BCT/OSUT/ UNIT	AN
N/A	Engage Targets with your assigned weapon	BCT/OSUT	AN
071-COM-0513	Select Hasty Fighting Positions (Repeat)	BCT/OSUT	SA

Army Warrior Training Plan			
Task Number	Title	Training Location	Sustainment Training Frequency
113-COM-1022	Perform Voice Communications (Repeat)	BCT/OSUT	AN
071-COM-0801	Challenge Persons Entering Your Area	BCT/OSUT	AN
113-COM-2070	Operate SINCGARS Single-Channel (SC)	BCT/OSUT	AN
071-COM-1004	Perform Duty as a Guard	BCT/OSUT	AN
071-COM-0815	Practice Noise, Light, and Litter Discipline (Repeat)	BCT/OSUT	AN
071-COM-0608	Use Visual Signaling Techniques (Repeat)	BCT/OSUT	AN
071-COM-4080	Send a SPOT Report	BCT/OSUT	AN
071-COM-4079	Send a SITREP Report	BCT/OSUT	AN
051-COM-1361	Camouflage Yourself and Individual Equipment	BCT/OSUT	AN
071-COM-4408	Construct an Individual Fighting Position	BCT/OSUT	AN
Battle Drill 3	Battle Drill Perform Tactical Combat Casualty Care	BCT/OSUT/ UNIT	AN
081-COM-0101	Evaluate a Casualty (Repeat)	BCT/OSUT	SA
081-COM-1005	Perform First Aid to Prevent or Control Shock (Repeat)	BCT/OSUT	SA
081-COM-1023	Open an Airway	BCT/OSUT	AN
081-COM-1007	Perform First Aid for Burns	BCT/OSUT	AN
081-COM-1054	Apply an Emergency Bandage	BCT/OSUT	AN
081-COM-0099	Apply a Hemostatic Dressing	BCT/OSUT	AN
081-COM-0069	Apply a Occlusive Dressing	BCT/OSUT	AN
081-COM-0048	Apply a Combat Application Tourniquet (CAT)	BCT/OSUT	AN
081-COM-1055	Apply a Fox Eye Shield	BCT/OSUT	AN

Army Warrior Training Plan			
Task Number	**Title**	**Training Location**	**Sustainment Training Frequency**
081-COM-0013	Initiate a DD Form 1380 Tactical Combat Casualty Care (TCCC) Card	BCT/OSUT	AN
081-COM-1046	Transport a Casualty (Repeat)	BCT/OSUT	AN
113-COM-1022	Perform Voice Communications (Repeat)	BCT/OSUT	AN
191-COM-0008	Search an Individual in a Tactical Environment	BCT/OSUT	AN
081-COM-1054	Evacuate casualties	BCT/OSUT	AN
Battle Drill 4	Battle Drill React to ambush (near)	BCT/OSUT/ UNIT	AN
N/A	Engage Targets with your assigned weapon	BCT/OSUT	AN
052-COM-1271	Identify Visual Indicators of an IED (Repeat)	BCT/OSUT	AN
052-COM-3261	React to an IED Attack (Repeat)	BCT/OSUT	AN
071-COM-0512	React-to-Hand-to-Hand-Combat (Repeat)	BCT/OSUT	AN
071-COM-4407	Employ Hand Grenades (Repeat)	BCT/OSUT	AN
071-COM-0501	Move as a member of a team (Repeat)	BCT/OSUT	AN
071-COM-0502	Move under direct fire (Repeat)	BCT/OSUT	AN
071-COM-0513	Select Hasty fighting positions (Repeat)	BCT/OSUT	AN
071-COM-0608	Use visual Signaling Techniques (Repeat)	BCT/OSUT	AN
113-COM-1022	Perform voice communication (Repeat)	BCT/OSUT	AN
Battle Drill 4	Battle Drill React to ambush (far)	BCT/OSUT/ UNIT	AN
N/A	Engage Targets with your assigned weapon	BCT/OSUT	AN
052-COM-1270	React to Possible Improvised Explosive Device (Repeat)	BCT/OSUT	AN
071-COM-0501	Move as a Member of a Team	BCT/OSUT	SA

Army Warrior Training Plan			
Task Number	*Title*	*Training Location*	*Sustainment Training Frequency*
071-COM-0513	Select Hasty Fighting Positions (Repeat)	BCT/OSUT	SA
113-COM-1022	Perform Voice Communications (Repeat)	BCT/OSUT	AN
071-COM-0608	Use Visual Signaling Techniques (Repeat)	BCT/OSUT	AN
071-COM-0510	React to Indirect Fire while Dismounted (Repeat)	BCT/OSUT	AN

THIS PAGE INTENTIONALLY LEFT BLANK

CHAPTER 3

Warrior Skills Level 1 Tasks

Subject Area 1: Shoot/Maintain, Employ, and Engage Targets with Individually Assigned Weapon System

071-COM-0032

Maintain an M16 Series Rifle/M4 Series Rifle Carbine

Foreign Disclosure: FD7 - This product/publication has been reviewed by the training/educational developers in coordination with the DOTD, MCoE, Ft Benning, GA 31905 FD authority. This product is NOT releasable to students from foreign countries.

WARNING

Do not squeeze the trigger until the weapon has been cleared. Inspect the chamber to ensure that it is empty and no ammunition is in position to be chambered. Failure to do so may lead to death or serius injury.

Conditions:
You have just returned from a mission with your loaded M16 series rifle or M4 series carbine and have been directed to conduct maintenance on your weapon. You have a small-arms accessory case. Some iterations of this task should be performed in MOPP 4.

Standards: Clear, disassemble, clean, inspect, lubricate, assemble, and perform a function check on the M16/M4. Maintain the magazine and ammunition.

Special Condition: None

Safety Risk: None

MOPP 4: Sometimes

Task Statements

Cue: None

DANGER

Do not squeeze the trigger until the weapon has been cleared. Inspect the chamber to ensure that it is empty and no ammunition is in position to be chambered. Failure to do so may lead to death or serious injury.

Note: None

Performance Steps

WARNING

Weapon must be cleared to be considered safe.

1. Clear the weapon.
   a. Point weapon in safe direction.
   b. Attempt to place the selector lever on SAFE.
Note: If weapon is not cocked, lever can't be pointed toward safe.
   c. Remove the magazine from the weapon, if present.
   d. Lock the bolt open.
      (1) Pull the charging handle rearward.
      (2) Press the bottom of the bolt catch.
      (3) Move the bolt forward until it engages the bolt catch.
      (4) Return the charging handle to the forward position.
      (5) Ensure the receiver and chamber are free of ammo.
   e. Place the selector lever on safe.
   f. Press the upper portion of the bolt catch to allow the bolt to go forward.
   g. Place the selector lever from SAFE to SEMI.
   h. Squeeze trigger.
   i. Pull the charging handle fully rearward and release it, allowing the bolt to return to the full forward position.
   j. Place the selector lever on SAFE.
2. Disassemble the weapon.
   a. Remove the sling.

CAUTION

Do not use a screwdriver or any other tool when removing the handguards. Doing so may damage the handguards, slip ring, or both.

Do not bend or dent the gas tube while removing handguard.

   b. Remove the handguards only if you can see dirt or corrosion through the vent holes.
Note: Hand guards on the M16A2 are interchangeable because they are identical. On the M16A4 the hand guards can be replaced by the M5 adapter rails. On the M4 carbine series, the hand guards can be replaced by the M4 adapter rails. The

M4 and M5 adapter rails are marked with a T for top and B for bottom. The operator is only authorized to remove the lower adapter rail and rail covers for cleaning, lubrication, or attaching accessories.

      (1) Place the weapon on the buttstock.

      (2) Press down on the slip ring with both hands.

      (3) Pull the handguards free.

   c. Push the take down pin as far as it will go.

   d. Pivot the upper receiver from the lower receiver.

   e. Push the receiver pivot pin in as far as it will go.

   f. Separate the upper and lower receivers.

   g. Remove carrying handle, if applicable.

      (1) Loosen the screws on the left side of the clamping bar.

      (2) Lift the handle off once the clamping bar is loose.

   h. Pull back the charging handle.

   i. Remove the bolt carrier and bolt.

   j. Remove the charging handle.

   k. Disassemble the bolt carrier.

      (1) Remove the firing pin retaining pin.

Note: Do not spread open or close split end of pin.

      (2) Push in bolt assembly to locked position.

CAUTION

Do not drop or hit the firing pin. Damage to the pin may cause the weapon to malfunction.

---

      (3) Drop firing pin out of rear of bolt carrier.

      (4) Remove the bolt cam pin by turning it one-quarter of a turn and lifting it out.

      (5) Remove bolt assembly from carrier.

      (6) Press the rear of the extractor pin to check spring function.

Note: Any weak springs should be reported to the unit armor for replacement.

      (7) Remove the extractor pin by pushing it out with the firing pin.

      (8) Lift out the extractor and spring, taking care that the spring does not separate from the extractor.

   l. Remove buffer and buffer spring from buttstock.

      (1) Press in buffer depress retainer and release buffer.

      (2) Remove buffer and action spring.

   m. Remove the buttstock. (M4 series only)

      (1) Extend the buttstock assembly to full open.

      (2) Separate the buttstock assembly from the lower receiver extension.

         (a) Grasp the lock lever in the area of the retaining nut.

         (b) Pull downward.

         (c) Slide the buttstock to the rear.

  3. Clean the weapon.

Note: CLP is used to identify when lubricant is needed, however it can be replaced with LSA (weapons lubricant oil, semifluid), or LAW (lubricating oil, arctic weather) as applicable.

Do not mix lubricants on the same weapon. The weapon must be thoroughly cleaned using dry cleaning solvent (SD) when changing from one lubricant to another.

CAUTION

Do not mix parts of one weapon with other weapons. Parts are not interchangeable.

---

    a. Clean the bore.

Note: The bore of your weapon has lands and grooves called rifling. Rifling makes the bullet spin very fast as it moves down the bore and down range. Because it twists so quickly, it is difficult to push a new, stiff bore brush through the bore. You will find it easier to pull your bore brush through the bore. Also, because the brush will clean better if the bristles follow the grooves (called tracking), you want the bore brush to be allowed to turn as you pull it through.

    (1) Attach three cleaning rod sections together.

    (2) Swab out the bore with a patch moistened with CLP or RBC.

    (3) Attach the bore brush.

Note: When using bore brush, don't reverse direction while in bore.

    (4) Point muzzle down.

    (5) Hold the upper receiver in one hand while inserting the end of the rod without the brush into the chamber.

    (6) Let the rod fall straight through the bore.

Note: About 2 to 3 inches will be sticking out of the muzzle at this point.

    (7) Attach the handle section of the cleaning rod to the end of the rod sticking out of the muzzle.

    (8) Pull the brush through the bore and out of the muzzle.

    (9) Take off the handle section.

    (10) Run the brush through the bore again by repeating the process.

    (11) Replace the bore brush with the rod tip.

    (12) Attach a patch with CLP to the rod tip.

    (13) Pull the patch through the bore.

    b. Upper receiver group.

    (1) Connect chamber brush to cleaning rod handle.

    (2) Dip the chamber brush in CLP and insert in chamber and locking lugs.

    (3) Push and twist to clean.

    (4) Use a worn out bore brush to clean outside of gas tube.

Note: Gas tubes will discolor from heat. Do not attempt to remove discoloration.

    (5) Clean the entire upper receiver by wiping it down.

    c. Bolt carrier group.

    (1) Clean carbon and oil from firing pin.

. (2) Clean bolt carrier key with worn brush.

(3) Clean firing pin recess with pipe cleaner.

(4) Clean firing pin hole with pipe cleaner.

(5) Clean behind bolt rings and lip of extractor.

(6) Clean carbon deposits and dirt from locking lugs.

CAUTION

Do not use wire brush or any other type of abrasive material to clean aluminum surfaces. Damage to equipment may occur.

---

d. Lower receiver group.

(1) Wipe dirt from trigger with a patch.

(2) Use a patch dampened with CLP to clean powder fouling, corrosion, and dirt from outside parts of lower receiver and extension assembly.

(3) Use pipe cleaner to clean buttstock drain hole.

(4) Clean buffer assembly, spring, and inside with patch dampened with CLP.

(5) Wipe dry.

e. Clean the ejector.

WARNING

Do not use a live round to perform this process.

---

(1) Place a few drops of CLP on the ejector.

(2) Press the ejector in using a spent round casing or dummy round.

(3) Hook casing under extractor and rock back and forth against ejector.

(4) Repeat this process a few times adding lubricant until the action of the ejector is smooth and strong.

(5) Dry off excess CLP when process is completed.

WARNING

Do not interchange bolts between weapons.

---

4. Inspect the weapon for serviceability.

a. Upper receiver group.

(1) Check handguards or rails for cracks, broken tabs, proper installation, and loose heat shields.

(2) Check front sight post for straightness.

(3) Check depression of the front detent.

(4) Check compensator for looseness.

(5) Check barrel for straightness, cracks, burrs or looseness.

(6) Check charging handle for cracks, bends, or breaks.

(7) Check rear sight assembly for properly working windage and elevation adjustments.

(8) Ensure the short and long range sight spring holds the selected sight in place.

(9) Check gas tube for bends or retention to barrel.

b. Bolt carrier group.

(1) Inspect bolt cam pin for cracking or chipping.

(2) Inspect firing pin for bends, cracks, and sharp or blunted tip.

Note: Bolts that contain pits in the firing pin hole need replacing.

(3) Inspect for missing or broken gas rings.

(4) Inspect bolt cam pin area for cracking or chipping.

(5) Inspect locking lugs for cracking or chipping.

(6) Inspect extractor assembly for missing extractor spring assembly with insert and for chipped or broken edges on the lip which engages the cartridge rim.

(7) Inspect firing pin retaining pin to determine if bent or badly worn.

(8) Inspect bolt carrier for loose bolt carrier key.

(9) Inspect for cracking or chipping in cam pin hole area.

c. Lower receiver.

(1) Inspect buffer for cracks or damage.

(2) Inspect buffer spring for kinks.

(3) Inspect buttstock for broken buttplate or cracks.

(4) Inspect for bent or broken selector lever.

(5) Inspect rifle grips for cracks or damage.

(6) Inspect for broken or bent trigger.

(7) Visually inspect the inside parts of the lower receiver for broken or missing parts.

d. Turn in weapons with unserviceable parts for maintenance.

5. Lubricate the weapon.

Note: Under all but the coldest arctic conditions, CLP is the lubricant to use on the weapon. Temperature between +10 degrees fahrenheit and -10 degrees fahrenheit, use either CLP or LAW. For -35 degrees fahrenheit or lower, use LAW only. Lightly lube means apply a film of lubricant barely visible to the eye. Generously lube means apply the lubricant heavily enough so that it can be spread with the finger.

a. Upper receiver and carrying handle.

(1) Lightly lubricate inside of upper receiver, bore, chamber, front sight, outer surfaces of barrel, and under the handguards.

(2) Apply a drop or two of lubricant to the front sight detent.

(a) Depress and apply two or three drops of CLP to the front sight detent.

(b) Depress several times to work the lube into the spring.

(3) Apply a drop or two of lubricant to both threaded studs.

(a) Lightly lube the clamping bar and both round nuts.

(b) Lightly lube the mating surface.

(4) Apply one or two drops of lubricant to the adjustable rear sight.

(5) Ensure that the lubricant is spread evenly in the rear sight by rotating the following parts.

(a) Elevation screw shaft.

(b) Elevation knob.

(c) Windage knob.

(d) Windage screw.

b. Lower receiver group.

(1) Lightly lube the inside and outside lower receiver extension, buffer, and action spring.

(a) Lightly lube the inside buttstock assembly.

(b) Generously lube the buttstock lock-release lever and retaining pin.

(2) Generously lube the take down pin, pivot pin, detents, and all other moving parts and their pins.

c. Bolt carrier group.

(1) Lightly lube the charging handle and the inner and outer surfaces of the bolt carrier.

(2) Place one drop of CLP in the carrier key.

(3) Apply a light coat of CLP on the firing pin and firing pin recess in the bolt.

(4) Generously lube the outside of the bolt body, bolt rings, and cam pin area.

(5) Apply a light coat of CLP on the extractor and pin.

6. Assemble the weapon.

a. Install the buttstock assembly. (M4 series only)

(1) Align the buttstock assembly with the lower receiver extension.

(2) Pull downward on the lock release lever near the retaining pin.

(3) Slide the buttstock assembly onto the lower receiver extension.

b. Insert the action spring and buffer.

c. Assemble the bolt carrier.

(1) Insert the extractor and spring.

(2) Push in the extractor pin.

(3) Slide the bolt into the carrier.

DANGER

The cam pin must be installed in the bolt group. Failure to do so will cause weapon to explode when fired next. Injury or death may occur.

---

(4) Replace the bolt cam pin.

(5) Drop in and seat the firing pin.

(6) Pull the bolt back.

(7) Replace the retaining pin.

d. Engage and then push the charging handle in part of the way.

e. Slide in the bolt carrier assembly.

f. Push in the charging handle and the bolt carrier group together.

g. Join the upper and lower receivers.

h. Engage the receiver pivot pin.

      i. Close the upper and lower receiver groups.

      j. Push in the take down pin.

      k. Replace the handguards.

      l. Replace the carrying handle, if applicable.

      m. Replace the sling.

    7. Perform a function check on the weapon.

    8. Maintain the magazine.

      a. Disassemble magazine.

        (1) Insert the nose of a cartridge into the hole in the base of the magazine.

        (2) Raise the rear of the magazine until the indentation on the base is clear of the magazine.

        (3) Slide the base forward until it is free of the tabs.

        (4) Remove the magazine spring and follower (do not separate).

      b. Clean all parts using a rag soaked with CLP.

      c. Dry all parts.

      d. Inspect parts for damage such as dents and corrosion.

Note: If any damage is found, turn in to maintenance.

      e. Lightly lube the spring only.

        (1) Insert the follower and spring into the magazine tube.

        (2) Jiggle the spring to seat them in the magazine.

        (3) Slide the base under all four tabs until it is fully seated.

        (4) Make sure the printing is on the outside.

      f. Assemble the magazine.

    9. Maintain the ammunition.

      a. Clean the ammunition with a clean dry rag.

      b. Inspect for and turn in any ammunition with the following defects:

        (1) Corrosion.

        (2) Dented cartridges.

        (3) Cartridges with loose bullets.

        (4) Cartridges with the bullet pushed in.

Evaluation Preparation:

Setup: Provide the Soldier with the equipment and or materials described in the conditions statement.

Brief Soldier: Tell the Soldier what is expected of him by reviewing the task standards. Stress to the Soldier the importance of observing all cautions, warnings, and dangers to avoid injury to personnel and, if applicable, damage to equipment.

Performance Measures	**GO**	**NO GO**
1 Cleared the weapon.	_____	_____

Performance Measures	GO	NO GO
2 Disassembled the weapon.	_____	_____
3 Cleaned the weapon.	_____	_____
4 Inspected the weapon for serviceability.	_____	_____
5 Lubricated the weapon.	_____	_____
6 Assembled the weapon.	_____	_____
7 Performed a function check on the weapon.	_____	_____
8 Maintained the magazine.	_____	_____
9 Maintained the ammunition.	_____	_____

Evaluation Guidance: Score the Soldier GO if all performance measures are passed. Score the Soldier NO-GO if any performance measure is failed. If the Soldier scores a NO-GO, show the Soldier what was done wrong and how to do it correctly.

References:
Required: TC 3-22.9, TM 9-1005-319-10
Related:

Environment: Environmental protection is not just the law but the right thing to do. It is a continual process and starts with deliberate planning. Always be alert to ways to protect our environment during training and missions. In doing so, you will contribute to the sustainment of our training resources while protecting people and the environment from harmful effects. Refer to ATP 3-34.5 Environmental Considerations and GTA 05-08-002 ENVIRONMENTAL-RELATED RISK ASSESSMENT.

References:
Safety: In a training environment, leaders must perform a risk assessment in accordance with ATP 5-19, Risk Management. Leaders will complete the current Deliberate Risk Assessment Worksheet in accordance with the TRADOC Safety Officer during the planning and completion of each task and sub-task by assessing mission, enemy, terrain and weather, troops and support available-time available and civil considerations, (METT-TC). *Note:* During MOPP training, leaders must ensure personnel are monitored for potential heat injury. Local policies and procedures must be followed during times of increased heat category in order to avoid heat related injury. Consider the MOPP work/rest cycles and water replacement guidelines IAW **TM 3-11.32**, Multi-Service Reference For Chemical, Biological, Radiological, And Nuclear Warning and Reporting And Hazard Prediction Procedures.

071-COM-0029

Perform a Function Check on an M16-Series Rifle/M4-Series Carbine.

Foreign Disclosure: FD6 - This product/publication has been reviewed by the product developers in coordination with the DOTD, MCoE, Ft Benning, GA 31905 foreign disclosure authority. This product is releasable to students from foreign countries on a case-by-case basis.

WARNING
Before starting functional check, be sure to clear the weapon. DO NOT squeeze the trigger until the weapon has been cleared. Inspect the chamber to ensure that it is empty and no ammunition is in position to be chambered.

Conditions: You are a member of a squad or team preparing for an tactical operation and must ensure the operability of your assigned M16-series rifle or M4-series carbine.

Standards: Conduct a function check and ensure that the weapon operates properly with the selector switch in each position.

Condition: None

Special Standards: None

Safety Risk: Low

Task Statements

Cue: None

Note: A function check is the final step of maintaining your weapon. It is also performed anytime the proper operation of a weapon is in question. Stop a function check at anytime the weapon does not function properly and turn in the malfunctioning weapon as per unit Standing Operating Procedures.

Performance Steps
   1. Confirm the M16/M4 is clear.
   2. Conduct a function check on the M16/M4.
     a. Place selector lever on SAFE.
     b. Pull charging handle to rear and release.
     c. Pull trigger.
Note: Hammer should not fall.
     d. Place selector lever on SEMI.
     e. Pull trigger.
Note: Hammer should fall.
     f. Hold trigger to the rear and charge the weapon.
     g. Release the trigger with a slow, smooth motion, until the trigger is fully forward.
Note: An audible click should be heard.
     h. Pull trigger.
Note: Hammer should fall.
     i. Place selector lever on BURST (M16A2, M16A4, and M4 only).
     j. Charge weapon one time.
     k. Squeeze trigger.
Note: Hammer should fall.
     l. Hold trigger to the rear.
     m. Charge weapon three times.
     n. Release trigger.
     o. Squeeze trigger.
Note: Hammer should fall.
     p. Place the selector switch on AUTO (M16A3 and M4A1 only).
     q. Pull the charging handle to the rear, charging the weapon.
     r. Squeeze the trigger.
Note: Hammer should fall.
     s. Hold the trigger to the rear.
     t. Cock the weapon again.
     u. Fully release the trigger then squeeze it again.
Note: The hammer should not fall because it should have fallen when the bolt was allowed to move forward during the chambering and locking sequences.

Evaluation Preparation:

Setup: Provide the Soldier with the equipment and/or materials described in the conditions statement.

Brief Soldier: Tell the Soldier what is expected by reviewing the task standards. Stress to the Soldier the importance of observing all cautions, warnings, and dangers to avoid injury to personnel and, if applicable, damage to equipment.

Performance Measures	GO	NO GO
1 Confirmed the M16/M4 was clear.	_____	_____
2 Conducted a function check on the M16/M4.	_____	_____

Evaluation Guidance: Score the Soldier GO if all performance measures are passed. Score the Soldier NO-GO if any performance measure is failed. If the Soldier scores a NO-GO, show the Soldier what was done wrong and how to do it correctly.

References:
Required:
Related: TC 3-22.9, TM 9-1005-319-10

071-COM-0028

Load an M16-Series Rifle / M4-Series Carbine

Foreign Disclosure: FD7 - This product/publication has been reviewed by the product developers in coordination with the DOTD, MCoE, Ft Benning, GA 31905 foreign disclosure authority. This product is NOT releasable to students from foreign countries.

Conditions: You are assigned a M16-series rifle or M4-series carbine and must load it in preparation for operation. You have a 5.56-mm ammunition in a 20 or 30 round magazine. Some iterations of this task should be performed in MOPP 4.

Standards: Keep the weapon pointed in a safe direction, ensure chanmber is empty, place weapon on safe, insert a magazine, and chamber a round.

Special Condition: None

Safety Risk: Medium

MOPP 4: Sometimes

Task Statements

Cue: None

Note: None

Performance Steps

1. Keep the weapon pointed in a safe direction..
2. Ensure chamber is clear.
   a. Pull the charging handle to the rear.
   b. Check the chamber to ensure it is clear.

Note: The chamber can be checked either by locking the bolt to the rear or by holding the bolt to the rear and then observing the chamber area.

   c. Return the charging handle to the forward position.
3. Place weapon on safe.
4. Lock the bolt to the rear.
   a. Pull the charging handle rearward.
   b. Press and hold the bottom of the bolt catch.
   c. Allow the bolt to move forward until it engages the bolt catch.
   d. Release the bottom of the bolt catch.
   e. Return the charging handle to the forward position.
5. Insert the magazine.

Note: Round may be chambered with the bolt assembly open or closed.

   a. Rotate the BURST cam to the BURST position.
   b. Ensure the bolt is forward and the selector level is on BURST.
6. Chamber a round.
   a. Chamber a round when the bolt is open.

Note: The charging handle should not be ridden forward.

      (1) Press the upper portion of the bolt catch allowing the bolt to go forward
      (2) Tap the forward assist to ensure that the bolt is fully forward and locked.

Note: The weapon is now loaded.

   b. Chamber a round when the bolt is closed.

Note: The charging handle should not be ridden forward.

      (1) Pull the charging handle to the rear as far as it will go.
      (2) Release the charging handle.
      (3) Tap the forward assist to ensure that the bolt is fully forward and locked.

Note: The charging handle should not be rode forward.

(Asterisks indicates a leader performance step.)

WARNING
The weapon is now loaded and should be pointed in a safe direction.

Evaluation Guidance: Score the Soldier GO if all performance measures are passed. Score the Soldier NO-GO if any performance measure is failed. If the Soldier scores a NO-GO, show the Soldier what was done wrong and how to do it correctly.

Evaluation Preparation: SETUP: Provide the Soldier with the equipment and/or materials described in the conditions statement.

Evaluation Preparation:

Setup: Provide the Soldier with the equipment and/or materials described in the conditions statement.

Brief Soldier: Tell the Soldier what is expected by reviewing the task standards. Stress to the Soldier the importance of observing all cautions, warnings, and dangers to avoid injury to personnel and, if applicable, damage to equipment.

Performance Measures	GO	NO GO
1 Keep the weapon pointed in a safe direction.	___	___
2 Ensured the chamber was clear.	___	___
3 Place the weapon on safe.	___	___
4 Locked the bolt to the rear.	___	___
5 Inserted a magazine.	___	___
6 Chambered a round.	___	___

Evaluation Guidance: Score the Soldier GO if all performance measures are passed. Score the Soldier NO-GO if any performance measure is failed. If the Soldier scores NO-GO, show the Soldier what was done wrong and how to do it correctly.

References:
Required: TC 3-22.9, TM 9-1005-319-10
Related:

071-COM-0027

Unload an M16-Series Rifle / M4-Series Carbine

Foreign Disclosure: FD7 - This product/publication has been reviewed by the training/educational developers in coordination with the DOTD, MCoE, Ft Benning, GA 31905 FD authority. This product is NOT releasable to students from foreign countries.

Conditions: You have just returned from a mission and have been directed to unload your M16-series rifle or M4-series carbine.Some iterations of this task should be performed in MOPP 4.

Standards: Unload the M16-series rifle or M4-series carbine so that the magazine and all ammunition are removed from the weapon.

Special Condition: None

Special Standards: None

Safety Risk: Medium

MOPP 4: Sometimes

Task Statements

Cue: None

Note: None

Performance Steps
  1. Point the weapon muzzle in a safe direction.
  2. Place the selector lever on SAFE.

Note: If the weapon is not cocked, you cannot place the selector lever on SAFE.

   3. Remove the magazine.

   4. Lock the bolt open.

     a. Pull the charging handle to the rear.

     b. Press the bottom portion of the bolt catch, locking the bolt open.

     c. Return the charging handle to the forward position.

     d. Place the selector lever on SAFE.

Note: If the weapon was cocked before locking the bolt open then the selector lever should already be on SAFE.

   5. Ensure that no ammunition is in the receiver and chamber.

   6. Return the bolt to the closed position.

     a. Press the upper portion of the bolt catch allowing the bolt to go forward.

     b. Place selector lever on SEMI.

     c. Pull the trigger to release the pressure on the firing pin spring.

     d. Close the ejection port cover.

(Asterisks indicates a leader performance step.)

Evaluation Preparation:

Setup: At a test site, provide an M4 or M4A1 carbine loaded with dummy ammunition.

Brief Soldier: Tell the Soldier to unload the carbine.

Performance Measures	GO	NO GO
1 Pointed the weapon muzzle in a safe direction.	_____	_____
2 Placed the selector lever on SAFE.	_____	_____
3 Removed the magazine.	_____	_____
4 Locked the bolt open.	_____	_____
5 Ensured no ammunition was in the receiver and chamber.	_____	_____

Performance Measures	GO	NO GO
6 Returned the bolt to the closed position.	_____	_____

Evaluation Guidance: Score the Soldier GO if all performance measures are passed. Score the Soldier NO-GO if any performance measure is failed. If the Soldier scores NO-GO, show the Soldier what was done wrong and how to do it correctly.

References
Required: TC 3-22.9, TM 9-1005-319-10
Related:

071-COM-0033

Correct Malfunctions of an M16-Series Rifle / M4-Series Carbine

Foreign Disclosure: FD7 - This product/publication has been reviewed by the training/educational developers in coordination with the DOTD, MCoE, Ft Benning, GA 31905 FD authority. This product is NOT releasable to students from foreign countries.

Conditions: You have a stoppage while engaging targets with your M16-series rifle or M4-series carbine. Some iterations of this task should be performed in MOPP 4.

Standards: Perform immediate and/or remedial action so you can continue to engage targets.

Special Condition: None

Ssfety Risk: Medium

MOPP 4: Sometimes

Task Statements

Cue: None

Note: None

1. Perform immediate action.

Note: The key word "SPORTS" will help you remember the steps for immediate action in sequence; slap, pull, observe, release, tap, shoot.

WARNING

When slapping up on the magazine, be careful not to knock a round out of the magazine into the line of the bolt carrier.

---

    a. Slap upward on the magazine to ensure it is fully seated and that the magazine follower is not jammed..

Note: When slapping up on the magazine, be careful not to knock a round out of the magazine into the line of the bolt carrier.

    b. Pull the charging handle fully to the rear.

    c. Observe the ejection of a live round or expended cartridge.

Note: If a weapon fails to eject a cartridge, perform remedial action.

    d. Release the charging handle; do not ride the charging handle.

    e. Tap the forward assist to ensure that the bolt is closed.

    f. Squeeze the trigger and try to fire the rifle.

DANGER

If weapon stops firing with a live round in the chamber of a hot barrel, remove the round quickly. However, if you cannot remove it within 10 seconds, remove magazine and wait 15 minutes with the weapon pointed in a safe direction. This will avoid injury during possible cook-off. Always keep face away from the ejection port when clearing a hot chamber.

Note: Apply immediate action only once for a stoppage. If the rifle fails to fire a second time for the same malfunction remedial action should be performed.

2. Perform remedial action.

    a. Correct an obstructed chamber.

      (1) Attempt to place the weapon on safe

      (2) Remove the magazine.

      (3) Lock the charging handle to the rear.

      (4) Place the selector lever on SAFE, if not already done.

      (5) Visually inspect the chamber.

DANGER

DO NOT attempt to remove a round stuck in the barrel of a weapon; turn the weapon in to field maintenance.

---

      (6) Remove obstructions from the chamber by:

        (a) Angling the ejection port downward and shaking the rifle to remove cartridge.

        (b) Using a cleaning rod to push out a cartridge stuck in the chamber.

b. Correct a mechanical malfunction.
   (1) Clear the weapon.
   (2) Disassemble the weapon.
   (3) Inspect for dirty, corroded, missing, or broken parts.
   (4) Clean dirty or corroded parts.
   (5) Replace missing or broken parts.
   (6) Assemble the weapon.
   (7) Perform a function check.
(Asterisks indicates a leader performance step.)

Evaluation Guidance: Score the Soldier GO if all performance measures are passed. Score the Soldier NO-GO if any performance measure is failed. If the Soldier scores a NO-GO, show the Soldier what was done wrong and how to do it correctly.

Evaluation Preparation:

Setup: Provide the Soldier with the equipment and/or materials described in the conditions statement.

Brief Soldier: Tell the Soldier what is expected by reviewing the task standards. Stress to the Soldier the importance of observing all cautions, warnings, and dangers to avoid injury to personnel and, if applicable, damage to equipment.

Performance Measures	GO	NO GO
1 Performed immediate action.	_____	_____
2 Performed remedial action.	_____	_____

Evaluation Guidance: Score the Soldier GO if all performance measures are passed. Score the Soldier NO-GO if any performance measure is failed. If the Soldier scores NO-GO, show the Soldier what was done wrong and how to do it correctly.

References:
Required: TC 3-22.9, TM 9-1005-319-10
Related:

071-COM-0031

Zero an M16 Series Rifle / M4-Series Carbine

Foreign Disclosure: FD7 - This product/publication has been reviewed by the training/educational developers in coordination with the DOTD, MCoE, Ft Benning, GA 31905 FD authority. This product is NOT releasable to students from foreign countries.

Conditions: You are assigned an M16-series rifle or M4-series carbine and have been directed to zero the weapon. You have 18 rounds of 5.56-mm ammunition, the appropriate 25-meter zero target, and sandbags for support. Some iterations of this task should be performed in MOPP 4.

Standards: Fire the weapon and adjust the sights so that five out of six rounds in two consecutive shot groups strike within the 4-centimeter circle on the target using 18 rounds or less. Record your zero.

Special Condition: None

Safety Risk: Medium

MOPP 4: Sometimes

Task Statements

Cue: None

Note: None

Performance Steps

    1. Set either the mechanical zero or battlesight zero on your weapon.
Note: Mechanically zeroing the weapon is only necessary when the weapon zero is questionable, the weapon is newly assigned to the unit, or the weapon sights have been serviced.
    a. Determine whether to set a mechanical zero or the battlesight zero.
      (1) Set a mechanical zero if-
        (a) The weapon sights have been serviced.
        (b) The weapon is newly assigned to the unit.
        (c) The current zero on the weapon is questionable.
      (2) Set a battlesight zero if a mechanical zero is not required.
    b. Set a mechanical zero on your weapon.
      (1) Adjust the Front Sight.

(a) Move the front sightpost until the base of the front sightpost is flush with the front sightpost housing.

(b) (M16A1 only) Move the front sightpost, from the flush position, 11 clicks in the direction of UP.

(2) Adjust the Rear Sight (by weapon type).

(a) (M16A1 only) Turn the rear sight windage drum left until it stops.

(b) (M16A1 only) Turn the windage drum right 17 clicks to center it.

(c) (M16A2 / M16A3 / M16A4 / M4-series) Set rear apertures by positioning the apertures so the unmarked aperture is up and the 0-200 meter aperture is down.

(d) (M16A2 / M16A3 / M16A4 / M4-series) Set windage by turning the windage knob to align the index mark on the 0-200 meter aperture with the long center index line on the rear sight assembly.

(e) (M16A2 / M16A3) Set the elevation of the M16A2/A3 by turning the elevation knob counterclockwise until the rear sight assembly rests flush with the carrying handle and the 8 / 3 marking is aligned with the index line on the left side of the carrying handle.

(f) (M16A4 only) Turn the elevation knob counterclockwise until the rear sight assembly rests flush with the carrying handle and the 6 / 3 marking is aligned with the index line on the left side of the carrying handle.

(g) (M4-series only) Turn the elevation knob counterclockwise until the rear sight assembly rests flush with the detachable carrying handle and the 6 / 3 marking is aligned with the index line on the left side of the carrying handle.

c. Set a battlesight zero on your weapon.

Note: No changes are made to the front sight when setting a battlesight zero.

(1) (M16A1 only) Adjust Rear Sight by flipping the aperture to ensure the aperture marked "L" is visible.

(2) (M16A2 / M16A3 / M16A4 / M4-Series only) Adjust rear aperture by positioning the apertures so the unmarked aperture is up and the 0-200 meter aperture is down.

(3) (M16A2 / M16A3 / M16A4 / M4-Series only) Adjust windage by turning the windage knob to align the index mark on the 0-200 meter aperture with the long center index line on the rear sight assembly.

(4) (M16A2 / M16A3 only) Adjust elevation by-

(a) Turning the elevation knob counterclockwise until the rear sight assembly rests flush with the carrying handle and the 8 / 3 marking is aligned with the index line on the left side of the carrying handle.

(b) Turning the elevation knob one more click clockwise.

(5) (M16A4 only) Adjust elevation by-

(a) Turning the elevation knob counterclockwise until the rear sight assembly rests flush with the carrying handle and the 6 / 3 marking is aligned with the index line on the left side of the carrying handle.

(b) Turning the elevation knob two more clicks clockwise so the index line on the left side of the detachable carrying handle is aligned with the "Z" on the elevation knob.

(6) (M4-series only) Adjust elevation by turning the elevation knob counterclockwise until the rear sight assembly rests flush with the detachable

carrying handle and the 6 / 3 marking is aligned with the index line on the left side of the carrying handle.

2. Establish a correct sight picture.

   a. Confirm the correct 25-meter zero target is facing you.

   b. Assume a prone supported firing position.

   c. Align the sights.

      (1) Center the top of the front sight post in the center of the rear sight.

      (2) Visualize imaginary cross hairs in the center of the rear aperture so that the top of the front sight post touches the imaginary horizontal line and the front sight post bisects imaginary vertical line.

      (3) Verify the sight picture.

   d. Align the aiming point.

      (1) Aim at target center.

      (2) Position the top of the front sight post center mass of the scaled silhouette target.

      (3) Confirm that an imaginary vertical line drawn through the center of the front sight post splits the target.

      (4) Confirm that an imaginary horizontal line drawn through the top of the front sight post splits the target.

3. Establish a tight shot group.

Note: A tight shot group is 3 consecutive rounds within a 4 centimeter or less circle.

   a. Fire a three round shot group at the 25-meter zeroing target.

   b. Identify the shot group on the target.

   c. Repeat step 3a and step 3b until 2 consecutive 3 round shot groups fall within a 4 centimeter or less circle.

Note: If a tight shot group is not obtained after 18 rounds then remedial training must be done.

4. Adjust sights (if required) to obtain a zero.

Note: Do not adjust the sights your just fired shot groups meet the standard.

   a. Determine the necessary sight adjustments by identifying the center of the last fired shot group and identifying the adjustment to move this point to the center of the strike zone (zero offset).

Note: The numbered squares around the edges of the target each represent a click on the sight.

   b. Adjust Elevation.

Note: One click clockwise moves the strike of the bullet down one square, while one click counterclockwise moves the strike of the bullet up one square.

      (1) Find the horizontal line nearest the center of the shot group.

      (2) Follow the line either left or right to the nearest edge of the target.

      (3) Identify the number of clicks and the direction of adjustment shown at the edge of the target.

      (4) Adjust the front sight in the indicated direction by the appropriate number of clicks.

      (5) Record the adjustment made on the target.

   c. Adjust Windage.

Note: Three clicks counterclockwise moves the strike of the bullet left one square, while three clicks clockwise moves the strike of the bullet right one square.

(1) Find the vertical line (up and down) nearest the center of the shot group.

(2) Follow the line either up or down to the nearest edge of the target.

(3) Identify the number of clicks and the direction of adjustment shown at the edge of the target.

(4) Adjust the rear sight in the indicated direction by the appropriate number of clicks.

(5) Record the adjustment made on the target.

5. Establish a zero.

a. Fire a three round shot group at the 25-meter zeroing target.

b. Identify the location of the shot group on the target.

(1) Return to step 4, if 2 of 3 rounds do not strike within the strike zone / zero offset.

(2) Proceed to step 6 if 2 of 3 rounds strike within the strike zone / zero offset.

6. Confirm the zero.

Note: A zero is confirmed when 5 of 6 rounds land within the center 4 centimeter center circle or the zero offset circle.

a. Fire a three round shot group at the 25-meter zeroing target.

b. Identify the location of the shot group on the target.

(1) Return to step 4, if 2 of 3 rounds do not strike within the strike zone / zero offset.

(2) Cease fire if 2 of 3 rounds strike within the strike zone / zero offset (your zero is confirmed).

7. (M4-series only) Rotate the rear sight elevation knob counterclockwise (down) two clicks to the 300-meter setting.

8. Record your zero.

a. Compute your zero.

b. Write your zero on a piece of tape.

c. Attach the tape to your weapon.

Evaluation Preparation:

Setup: Provide the Soldier with the equipment and or materials described in the conditions statement.

Brief Soldier: Tell the Soldier what is expected of him by reviewing the task standards. Stress to the Soldier the importance of observing all cautions, warnings, and dangers to avoid injury to personnel and, if applicable, damage to equipment.

Performance Measures	GO	NO GO
1 Set either the mechanical zero or the battlesight zero on your weapon.	_____	_____
2 Established a correct sight picture.	_____	_____
3 Established a tight shot group.	_____	_____
4 Adjusted sights (if required) to obtain a zero.	_____	_____
5 Established a zero.	_____	_____
6 Confirmed the zero.	_____	_____
7 (M4-series only) Rotated the rear sight elevation knob counterclockwise (down) two clicks to the 300-meter setting.	_____	_____
8 Recorded your zero.	_____	_____

Evaluation Guidance: Score the Soldier GO if all performance measures are passed. Score the Soldier NO-GO if any performance measure is failed. If the Soldier scores a NO-GO, show the Soldier what was done wrong and how to do it correctly.

References
Required: TC 3-22.9, TM 9-1005-319-10
Related:

071-COM-0030

Engage Targets with an M16-Series Rifle/M4 Series Carbine

Foreign Disclosure: FD3 - This training product has been reviewed by the developers in coordination with the G2, Ft Benning, GA 31905 foreign disclosure officer. This training product cannot be used to instruct international military students.

Conditions: You are a member of a squad conducting dismounted operations and have been assigned a sector of fire by your leader. You have your M16 series rifle or M4 series carbine, magazines, ammunition, and individual combat/personal protective equipment. Some iterations of this task should be performed in MOPP 4.

Standards: Select a firing position and engage targets in your assigned sector until they no longer present a threat or you are directed to cease fire.

Special Conditions: None

Safety Risk: High

MOPP 4: Sometimes

Task Statements

Cue: None

Performance Steps

1. Select a position that allows for adequate observation of assigned sector of fi
Note: Your situation should affect your physical positioning and firing stance. Your position should protect you from enemy fire and observation, yet allow you to place effective fire on targets in your sector of fire. Your position may varyfrom a fixed location to a temporary location during movement.
Note: Detection of targets depends on your position, your skill in scanning, and your ability to observe the are and recognize target indicators.
2. Scan sector of fire using one of the following methods.
   a. Self preservation method

   b.50-meter overlapping strip method

   c.Maintaining observation on the area

3. Identify targets in designated sector of fire.

4. Determine range to targets

  a. 100-meter unit of measure method

  b. Appearance of objects method.

  c. Front sight post method
  d. Appearance of objects method.
  e. Combination method.
  5. Fire on targets using correct fundamentals of marksmanship and appropriate aiming and engagement techniques
  a. Apply the fundamentals of marksmanship.
  (1) Steady position
  (2) Aiming.
  (3) Breath control.
  (4) Trigger squeeze.
  b. Use appropriate aiming and engagement techniques as needed.
  (1) Combat fire techniques.
  (2) Chemical, biological, radiological and nuclear (CBRN) firing.
  (3) Night firing.
  (4) Moving targets.
  (5) Short-range marksmanship techniques.
  (6) Cease fire on targets once they are destroyed, suppressed, or you receive an order to cease fire.

Evaluation Preparation:

Setup: On a live-fire range, provide sufficient quantities of equipment and ammunition to support the number of Soldiers tested. Have each Soldier use his own rifle and magazine.

Brief Soldier: Tell the Soldier what is expected by reviewing the task standards. Stress to the Soldier the importance of observing all cautions, warnings, and dangers to avoid injury to personnel and, if applicable, damage to equipment.

	Performance Measures	GO	NO GO
1	Selected a position that allowed for adequate observation of assigned sector of fire.	____	____
2	Scanned sector of fire.	____	____
3	Identified target in designated sector of fire.	____	____

Performance Measures	GO	NO GO
4 Determined range to target.	_____	_____
5 Fired on targets.	_____	_____
6. Ceased fire once targets were destroyed, suppressed, or you were directed to cease fire .	_____	_____

Evaluation Guidance: Score the Soldier GO if all performance measures are passed. Score the Soldier NO-GO if any performance measure is failed. If the Soldier scores a NO-GO, show the Soldier what was done wrong and how to do it correctly.

References:
Required: TC 3-22.9, TC 3-21.75, TM 9-1005-319-10

Related:

Subject Area 2: Shoot/Employ Hand Grenades

071-COM-4401

Perform Safety Checks on Hand Grenades

Foreign Disclosure: FD7 - This product/publication has been reviewed by the training/educational developers in coordination with the DOTD, MCoE, Ft Benning, GA 31905 FD authority. This product is NOT releasable to students from foreign countries.

Conditions: You are a member of a squad or team preparing for a mission and have been directed to perform safety checks on the hand grenades issued to your squad/team. The hand grenades are in a shipping container. You are wearing your individual combat/personal protective equipment. Some iterations of this task should be performed in MOPP 4.

Standards: Inspect the shipping container, canister, and hand grenade for defects; report and turn in hand grenade that has defect(s) that cannot be corrected; secure hand grenade(s) properly in carrying pouch(s).

Special Condition: None

Safety Risk: Medium

MOPP 4: Somrtimes

Task Statements

Cue: None

*Note:*If any discrepancies are found upon receipt of an issued shipping container, canister or hand grenade, personnel should return the shipping container, canister or hand grenade to the issuing person or dispose of it in accordance with the unit tactical standing operating procedures(TACSOP).

Performance Steps

1. Inspect hand grenade shipping container (Figure 071-COM-4401-1), if applicable.

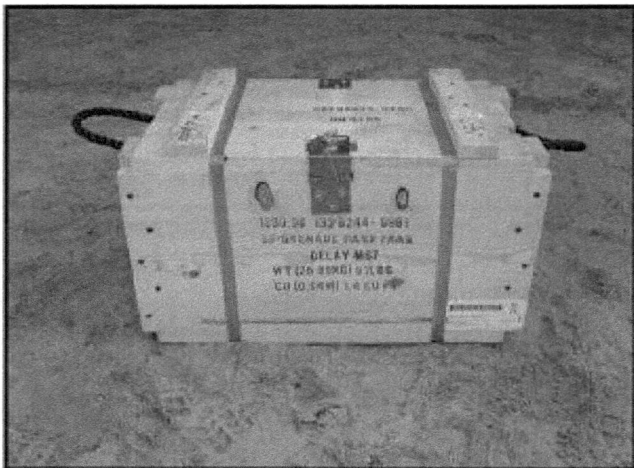

Figure 071-COM-4401-1 Shipping container.

   a. Shipping container is not damaged.
   b. Inform supervisor if shipping container is damaged.
2. Inspect the grenade canister (Figure 071-COM-4401-2), if applicable.

Figure 071-COM-4401-2. Grenade canister.

    a. Inspect the canister for damage.

    (1) Check to see if seal on the canister has been tampered with or is missing.

    (2) Ensure canister is not dented or punctured.

    (3) Inform supervisor of any deficiencies found.

    b. Open the canister.

WARNING

Do not attempt to remove the grenade found upside down in its packing container.

---

    (1) Check to see if the grenade is upside down inside of the shipping canister.

    (a) Replace canister top and tape in place if grenade found upside down.

    (b) Report deficiencies to supervisor.

    (c) Return canister to ammunition disposal personnel.

    (2) Check to see if the safety pin is in proper position.

    (a) Ensure that safety pin is in place and undamaged.

    (b) Check that the legs of the safety pin have either angular spread or diamond crimp.

    (3) Ensure safety clip (when installed) is in place and undamaged.

   3. Inspect the hand grenade.

    a. Remove the packing material and the hand grenade from the canister.

    (1) Check for rust on the body or the fuze.

    (2) Ensure holes are not visible in the body or the fuze.

    (3) Check hand grenade for cracked body.

    (4) Place back in canister if any defect(s) are found, if applicable.

    b. Ensure the safety pin (1) is secured properly (Figure 071-325-4401-3).

Note: If not properly secured, carefully push it into place while holding the safety lever down.

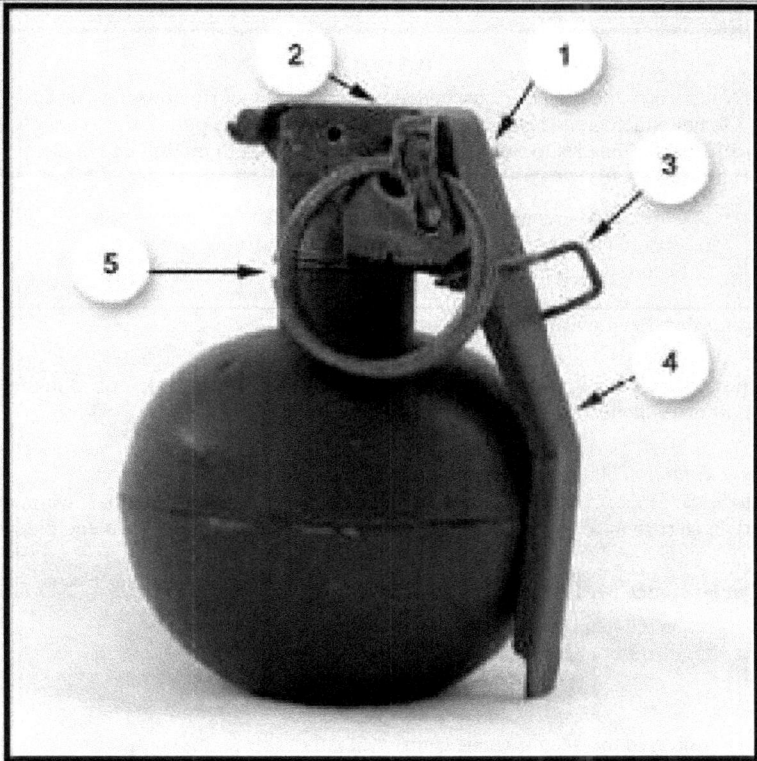

Figure 071-COM-4401-3. Grenade components.

    c. Ensure the confidence clip (2) is present and properly secured to the pull ring.

    d. Ensure the safety clip (3) is present and properly secured to the safety lever (4).

Note: If not properly secured, carefully push it into place while holding the safety lever down.

WARNING

Never remove the fuze from a live grenade.

---

    e. Check the hand grenade fuze assembly (5) for tightness.

    f. Ensure the safety lever (4) is not bent or broken.

    g. Turn in defective hand grenade, if applicable.

  4. Secure the grenade.

> **WARNING**
>
> Never carry the grenades suspended by the safety pull ring or safety lever. Do not attach grenades to clothing or equipment by the pull ring. Do not tape hand grenades to Soldier's gear. Do not attempt to modify a grenade.

    a. Carry hand grenades using the proper procedures.
    b. Ensure that the grenade is fully inside the carrying pouch.
    c. Secure pouch flap.

Evaluation Preparation:

Setup: Provide the Soldier with the equipment and or materials described in the conditions statement.

Brief Soldier: Tell the Soldier what is expected of him by reviewing the task standards. Stress to the Soldier the importance of observing all cautions, warnings, and dangers to avoid injury to personnel and, if applicable, damage to equipment.

Performance Measures	GO	NO GO
1 Inspected hand grenade shipping container, if applicable.	___	___
2 Inspected the hand grenade canister, if applicable.	___	___
3 Inspected the hand grenade.	___	___
4 Secured the hand grenade.	___	___
5 Report any deficiencies to Supervisor.	___	___

Evaluation Guidance: Score the Soldier GO if all performance measures are passed. Score the Soldier NO-GO if any performance measure is failed. If the Soldier scores a NO-GO, show the Soldier what was done wrong and how to do it correctly.

References:
Required: TC 3-23-30; TM 9-1330-200-12
Related:

071-COM-4407

Employ Hand Grenades

Foreign Disclosure: FD3 - This training product has been reviewed by the developers in coordination with the G2, Ft Benning, GA 31905 foreign disclosure officer. This training product cannot be used to instruct international military students.

Conditions: Given a fragmentation, concussion, riot control, smoke, or incendiary grenade with a time-delay fuse, a point or area target to engage, and load bearing vest (LBV), load bearing equipment (LBE), Modular, Lightweight, Load-bearing, Equipment (MOLLE) or Improved Outer Tactical Vest (IOTV). Some iterations of this task should be performed in MOPP 4.

Standards: Engage target with a hand grenade by: selecting appropriate hand grenade based on type target, determining throwing position, correctly gripping, preparing, and throwing the hand grenade so it is within the effective range of the target.

Special Condition: None

Special Standards: None

Safety Risk: High

Cue: None

MOPP 4: Sometimes

Task Statements

Performance Steps

    1. Select appropriate hand grenade based on type of target.
    2. Select proper throwing position.
Note: You can use five positions to throw grenades - standing, prone-to-standing, kneeling, prone-to-kneeling, and alternate prone. However, If you can achieve more distance and accuracy using your own personal style, do so as

long as your body is facing sideways and toward the enemy's position, and you throws the grenade overhand.

    a. Ensure you have a proper covered position.

    b. Determine the distance to the target.

    c. Align your body with the target.

  3. Grip the hand grenade.

Note: Do not remove the safety clip or the safety pin until the grenade is about to be thrown.

    a. Place the hand grenade in the palm of the throwing hand with the safety lever placed between the first and second joints of the thumb.

Note: For left handed throwers the grenade is inverted with the top of the fuze facing downwards in the throwing hand.

    b. Keep the pull ring away from the palm of the throwing hand so that it can be easily removed by the index or middle finger of the free hand.

  4. Prepare the hand grenade.

    a. Tilt the grenade forward to observe the safety clip.

    b. Remove the safety clip by sweeping it away from the grenade with the thumb of the opposite hand.

    c. Insert the index or middle finger of the nonthrowing hand in the pull ring until it reaches the knuckle of the finger (Figure 071-COM-4407-1).

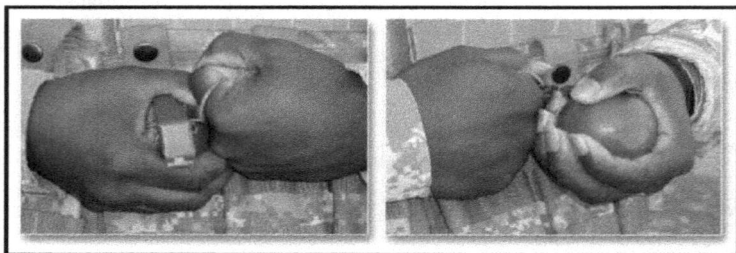

Figure 071-COM-4407-1. Pull ring grip, right/left hand.

> **DANGER**
>
> If pressure on the safety lever is relaxed after the safety clip pin are removed, the striker can rotate and strike the primer while the thrower is still holding the grenade. Continuing to hold the grenade beyond this point can result in injury or death.

    d. Ensure that you are holding the safety lever down firmly.

    e. Twist the pull ring toward the body (away from the body for left handed throwers) to release the pull ring from the confidence clip.

CAUTION

Never attempt to reinsert a safety pin into a hand grenade during training. In combat, however, it may be necessary to reinsert a safety pin into a grenade. Take special care to replace the pin properly. If the tactical situation allows, it is safer to throw the grenade rather than to trust the reinserted pin.

f. Remove the safety pin by pulling the pull ring from the grenade (Figure 071-COM-4407-2 and Figure 071-COM-4407-3).

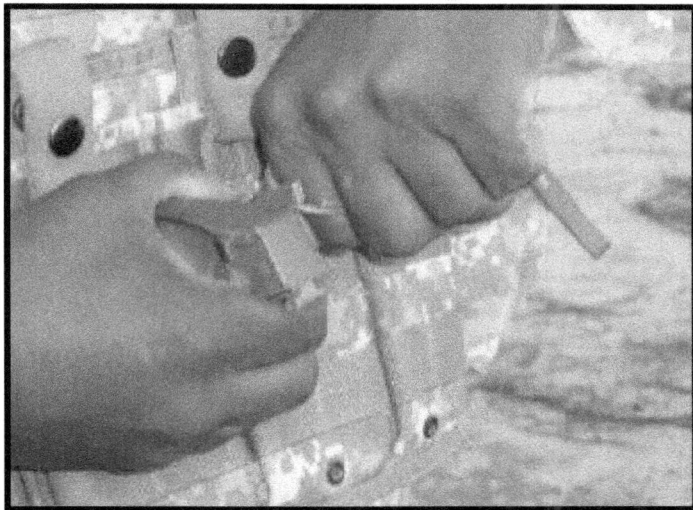

Figure 071-COM-4407-2. Right hand grip, pulling safety pin.

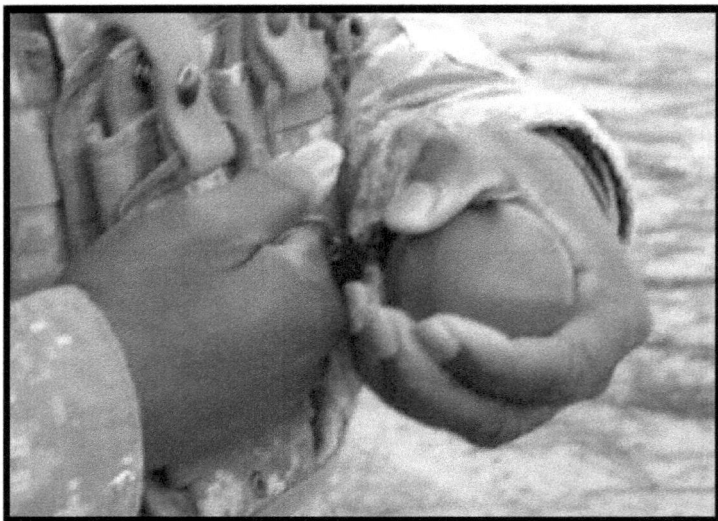

Figure 071-COM-4407-3. Left hand grip, pulling the safety pin.

5. Throw the hand grenade so it is within the effective range of the target.
   a. Observe the target to estimate the distance between the throwing position and the target area.

Note: In observing the target, minimize exposure time to the enemy (no more than 3 seconds).

WARNING

The flight path of the grenade must be checked to make sure no obstacles alter the flight of the grenade or cause it to bounce back toward you.

---

   b. Ensure there are no obstacles that can alter or block the flight of the grenade when it is thrown.
   c. Confirm body target alignment.

DANGER

Use cook-off procedures only in a combat environment. In training, never cook off live fragmentation hand grenades or offensive concussion grenades.

Never cook off the M84, stun grenade, or smoke grenades. These grenades have short fuze delays (1 to 2.3 seconds) and will cause serious personal injury if cook-off procedures are performed.

The grenade must be thrown immediately after count off.

    d. Cook off the hand grenade. (Optional)

Note: Cooking off uses enough of the grenade's 4- to 5-second delay (about 2 seconds) to cause the grenade to detonate above ground or shortly after impact with the target.

    (1) Release the safety lever.

    (2) Count "One thousand one, one thousand two".

    e. Throw the grenade overhand so that the grenade arcs, landing on or near the target.

Note: To be effective the target must be within the bursting radius of the grenade.

    f. Allow the motion of the throwing arm to continue naturally once the grenade is released.

    g. Seek cover to avoid being hit by fragments or direct enemy fire.

Note: If no cover is available, drop to the prone position with your protective head gear facing the direction of the grenade's detonation.

Evaluation Preparation:

Setup: Provide the Soldier with the equipment and or materials described in the conditions statement.

Brief Soldier: Tell the Soldier what is expected of him by reviewing the task standards. Stress to the Soldier the importance of observing all cautions, warnings, and dangers to avoid injury to personnel and, if applicable, damage to equipment.

Performance Measures	GO	NO GO
1 Selected the appropriate hand grenade based on type of target.	_____	_____
2 Selected appropriate throwing position.	_____	_____
3 Gripped the hand grenade.	_____	_____
4 Prepared the grenade.	_____	_____
5 Threw the hand grenade so it was within the effective range of the target.	_____	_____

Evaluation Guidance: Score the Soldier GO if all performance measures are passed. Score the Soldier NO-GO if any performance measure is failed. If the Soldier scores a NO-GO, show the Soldier what was done wrong and how to do it correctly.

References:
Required:
Related: TC 3-23.30, TM 9-1330-200-12

Subject Area 3: Move

071-COM-0541

Perform Exterior Movement Techniques during an Urban Operation

Foreign Disclosure: FD7 - This product/publication has been reviewed by the product developers in coordination with the DOTD, MCoE, Ft Benning, GA 31905 foreign disclosure authority. This product is NOT releasable to students from foreign countries.

Conditions: You are a member of a dismounted squad or team conducting movement within an urban area. You have your assigned weapon and individual/protective equipment. The enemy's location and strength in the area are unknown.

Standards: Move within an urban area using proper urban movement techniques while minimizing exposure to enemy fire.

Special Condition: None

Safety Risk: Medium

MOPP 4:

Task Statements

Cue: None

Note: Outdoor movement in urban terrain is best conducted as part of a buddy team, a fire team, or a squad. This ensures at least one Soldier is providing overwatch of another Soldier's movement, either from a stationary position or as both are moving, and prevents individual Soldiers from being isolated. This allows for a rapid engagement of any enemy that either exposes themselves (such as by leaning out of or by silhouetting themselves in a window) or by firing.

Performance Steps
1. Move across streets or open areas.
Note: Open areas include parks, plazas and large intersections as well as streets, open air buildings, and large rooms that are significantly exposed to exterior view. Ideally, avoid these open areas as they are potential killing zones for the enemy, especially crew-served weapons and snipers; however,

operations often require movement across these areas. Cross these areas using the same basic techniques used to cross any danger area.

    a. Identify the far side position before moving with a clear understanding of how it will be occupied or cleared.

    b. Conduct a visual reconnaissance of all the dimensions of urban terrain to identify likely threat positions.

    c. Select a position on the far side that provides the best available cover.

    d. Select the best route to the far side position that minimizes the time exposed.

Note: Obscurants, such as smoke, are an option to conceal movement. However, thermal sighting systems can see through smoke and when smoke is thrown in an open area, an enemy may fire into the smoke cloud in anticipation of movement through or behind the smoke.

    e. Cross rapidly along the selected route to the selected position.

    2. Move parallel to buildings.

Note: Moving parallel to buildings is the movement normally associated with moving down a roadway but also includes movement in plazas or other open areas that are between buildings. During contact, utilize smoke, suppressive fires, and individual movement techniques. In moving to adjacent buildings, team members should keep a distance of 3 to 5 meters between themselves, leapfrogging along each side of the street and from cover to cover.

    a. Soldier moves parallel to the side of a building.

    b. Use existing cover and concealment.

    c. Stay in the shadows.

    d. Present a low silhouette.

    e. Use proper techniques to cross door and window openings

    f. Move rapidly to the next position.

    3. Move past building opening (windows and doors).

Note: The most common mistakes at windows are exposing the head in a first-floor window and not being aware of basement windows.

    a. Move past an above-knee window.

      (1) Stay near the side of the building.

      (2) Stay below the level of the window.

      (3) Avoid silhouetting self in window (Figure 071-COM-0541-1).

Figure 071-COM-0541-1. Soldier moving past window.

    b. Move past a below-knee window (basement).
        (1) Stay near the side of the building.
        (2) Step or jump past the window without exposing legs (Figure 071-COM-0541-2).

Figure 071-COM-0541-2. Soldier moving past basement window.

c. Move past a full-height window (store type) or open door.

Note: A Soldier should not just walk past an adjacent full height window, as he presents a good target to an enemy inside the building.

(1) Identify a position on the far side of the window.

(2) Determine which technique to use to cross the opening.

(a) Run across the opening to the far side.

(b) Arc around the opening while covering the opening with your weapon while moving.

(3) Move rapidly to the far side position.

4. Move around corners.

Note: Before moving around a corner, the Solider must first observe around the corner. The most common two mistakes Soldier make at corners are exposing their head and upper body where it is expected and flagging their weapon.

a. Move around a corner by first observing around the corner.

(1) Lie flat on the ground, weapon at your side, ensuring that your weapon is not forward of the corner.

Note: DO not show your head below at the height an enemy would expect to see it.

(2) Expose your head (with Helmet) only enough to observe around the corner (Figure 071-COM-0541-3).

Note: When speed is required the Pie-ing method is applied.

Figure 071-COM-0541-3. Soldier looking around a corner.

(3) Continue movement around the corner, if clear.

b. Move around the corner by using the pie-ing method.

(1) Aim the weapon beyond the corner (without flagging) into the direction of travel.

(2) Side-step around the corner in a semi-circular fashion with the muzzle as the pivot point (Figure 071-COM-0541-4).

Figure 071-COM-0541-4. Soldier Pie-ing around a corner.

    (3) Continue movement around the corner, if clear.

  5. Cross a wall.

    a. Reconnoiter the other side.

Note: The far side must be relatively safe from enemy fire, as once across the wall, the Soldier is fully exposed. Additionally, the immediate opposite side of the wall must be safe for landing; long drops and debris can cause injury.

    b. Identify a far side position.

Note: Once across the wall, you will then move to this far side position. This position may be at the wall, near the wall, or away from the wall.

    c. Crouch near the wall.

    d. Hold your weapon with one hand while grabbing the top of the wall with the other hand

    e. Pull with the hand on the wall while simultaneously swinging both legs over the wall, one right after the other.

    f. Roll your whole body quickly over the wall while keeping a low silhouette (Figure 071-COM-0541-5).

Note: Speed of movement and a low silhouette deny the enemy a good target.

Figure 071-COM-0541-5. Soldier crossing a wall.

g. Move to you next position once on the far side.

Evaluation Preparation:

Setup: At the test site, provide all materials and equipment given in the task condition statement.

Brief Soldier: Tell the Soldier to move as a designated member of an assult element in urban terrain. The enemy strength and location are unknown.

Performance Measures	GO	NO GO
1 Moved across a street or open area.	_____	_____
2 Moved parallel to a building.	_____	_____
3 Moved passed a building opening (window or open door).	_____	_____

Performance Measures	GO	NO GO
4 Moved around a corner.	_____	_____
5 Crossed a wall.	_____	_____

Evaluation Guidance: Score the Soldier GO if all performance measures are passed. Score the Soldier NO-GO if any performance measure is failed. If the Soldier scores NO-GO, show the Soldier what was done wrong and how to do it correctly.

References:
Required: ATTP 3-06.11; TC 3-21.75
Related:

071-COM-0503

Move Over, Through, or Around Obstacles (Except Minefields)

Foreign Disclosure: FD7 - This product/publication has been reviewed by the training/educational developers in coordination with the DOTD, MCoE, Ft Benning, GA 31905 FD authority. This product is NOT releasable to students from foreign countries.

Conditions: As a member of a dismounted team conducting movement to contact, you encounter a natural or manmade obstacle. You have your assigned weapon and individual/protective equipment. The enemy's location and strength in the area are unknown. Some iterations of this task should be performed in MOPP 4.

Standards: Notify chain of command of obstacle encountered, evaluate obstacle, identify nearest covered position on far side of obstacle, negotiate a wall obstacle, and provide local security for follow on forces during engotiation or reduction of obstacle.

Special Condition: None

Special Standards: None

Safety Risk: Medium

MOPP 4: Sometimes

Task Statements

Cue: None

*Note:*An obstacle is any obstruction designed or employed to disrupt, fix, turn, or block the movement of an opposing force, and to impose additional losses in personnel, time, and equipment on the opposing force. Obstacles can be natural, manmade, or a combination of both.

Performance Steps

1. Notify your chain of command of the presence and type of obstacle encountered.
Note: Most obstacles, for maximum effectiveness, are covered by either fire or observation. Many obstacles, due to enemy fire or complexity of the obstacle, require a unit breaching operation and the appropriate collective task should also be followed.

2. Evaluate the obstacle, from a covered position, to determine whether to move around, through or over the obstacle
Note: Typically it is best to move around (or bypass) an obstacle, however this is not always possible.

3. Identify the nearest covered position on the far side of the obstacle.

4. Ensure a buddy, if present, covers your movement as you negotiate the obstacle.

5. Negotiate a wall obstacle.

   a. Identify your immediate landing position on the far side of the wall.
Note: The far side must be relatively safe from enemy fire, as once across the wall, you are fully exposed. Additionally, the immediate opposite side of the wall must be safe for landing as long drops and debris can cause injury.

   b. Assume a crouching position near the wall, while holding your weapon with one hand and grabbing the top of the wall with the other hand.

   c. Pull with the hand on the wall while simultaneously swinging both legs over the wall, one right after the other.

   d. Roll quickly over the top to other side, keeping a low silhouette.

   e. Move to the identified covered position on the far side.

WARNING
An enemy may attach booby traps or tripwire-activated mines to wire obstacles.
---

6. Negotiate a wire obstacle.

   a. Move to your designated crossing position.

   b. Check for booby traps or early warning devices.

    c. Cross over a wire obstacle.

       (1) Place an object such as a piece of wood, metal, or mats, over the wire.

       (2) Move over the wire by stepping on this object to avoid the wire entanglements.

    d. Cross under a wire obstacle.

       (1) Slide head first on your back under the bottom strands.

       (2) Push yourself forward with your shoulders and heels, carrying your weapon lengthwise on your body and holding the barbed wire with one hand while moving.

       (3) Let the barbed wire slide on the weapon to keep wire from catching on clothing and equipment.

    e. Cut through a wire obstacle.

Note: If stealth is not needed then quickly cut all wires and proceed through the gap.

       (1) Wrap cloth around the barbed wire between your hands.

       (2) Cut partly through the barbed wire.

Note: Cutting the wire near a picket reduces the noise of a cut.

       (3) Bend the barbed wire back and forth quietly until it separates.

       (4) Cut only the lower strands.

       (5) Cross under the remaining top wires.

   7. Cross a ditch type obstacle.

    a. Select a point that has cover and concealment on both sides, such as a bend in the ditch.

    b. Move to your designated crossing site.

    c. Crawl up to the edge of the open area.

    d. Observe both the floor of the ditch and the far side for dangers.

    e. Move rapidly but quietly across the exposed area.

    f. Assume a covered position on the far side.

   8. Cover your buddy, if present, as he or she crosses the obstacle.

Evaluation Preparation:

Setup: Provide the Soldier with the equipment and or materials described in the conditions statement.

Brief Soldier: Tell the Soldier what is expected of him by reviewing the task standards. Stress to the Soldier the importance of observing all cautions, warnings, and dangers to avoid injury to personnel and, if applicable, damage to equipment.

Performance Measures	GO	NO GO
1 Notified the chain of command of the presence and type of obstacle encountered.	_____	_____

Performance Measures	GO	NO GO
2 Evaluated the obstacle, from a covered position, to determine whether to move around, through or over the obstacle.	_____	_____
3 Identified the nearest covered position on the far side of the obstacle.	_____	_____
4 Ensured team members, if present, provide local security for your movements as you negotiated the obstacle.	_____	_____
5 Negotiated a wall obstacle.	_____	_____
6 Provide local security on far side of obstacles for follow on forces, if present.	_____	_____

Evaluation Guidance: Score the Soldier GO if all performance measures are passed. Score the Soldier NO-GO if any performance measure is failed. If the Soldier scores a NO-GO, show him what was done wrong and how to do it correctly.

References
Required: TC 3-21.75
Related:

071-COM-1000

Identify Topographic Symbols on a Military Map

Foreign Disclosure: FD7 - This product/publication has been reviewed by the product developers in coordination with the DOTD, MCoE, Ft Benning, GA 31905 foreign disclosure authority. This product is NOT releasable to students from foreign countries.

Conditions: You are a member of a squad or team in a field environment and have been given; a 1:50,000 scale military map and a requirement to identify topographic symbols on the map.

Standards: Identify topographic symbols, colors, and marginal information on a military map.

Special Condition: None

Special Standards: None

Safety Risk: Low

MOPP 4:

Task Statements

Cue: None

Note: None

Performance Steps

1. Identify the six basic colors on a military map (Figure 071-COM-1000-1).

COLORS	SYMBOLS
Black	Cultural (man-made) features other than roads
Blue	Water
Brown	All relief features - contour lines on old maps - cultivated land on red-light readable maps
Green	Vegetation
Red	Major roads, built-up areas, special features on old maps
Red-brown	All relief features and main roads on red-light readable maps

Figure 071-COM-1000-1. Colors

    a. Identify the features that the color black represents.
Note: Indicates cultural (manmade) features such as buildings and roads, surveyed spot elevations, and all labels.
    b. Identify the features that the color blue represents.
Note: Indicates hydrography or water features such as lakes, swamps, rivers, and drainage.
    c. Identify the features that the color green represents.
Note: Indicates vegetation with military significance such as woods, orchards, and vineyards.
    d. Identify the features that the color brown represents.
Note: Brown identifies all relief features and elevation such as contours on older edition maps and cultivated land on red light readable maps.
    e. Identify the features that the color red represents.
Note: Classifies cultural features, such as populated areas, main roads, and boundaries, on older maps.
    f. Identify the features that the color red-brown represents.

Note: These colors are combined to identify cultural features, all relief features, non surveyed spot elevations, and elevation such as contour lines on red light readable maps.

 g. Identify all other features and the colors they represent, if applicable.
Note: Other colors may be used to show special information. These are indicated in the marginal information as a rule.

 2. Identify the symbols on a military map.

  a. Use the legend, which should identify most of the symbols used on the map.

  b. Identify each object by its shape on the map.
Note: For example, a black, solid square represents a building or a house; a round or irregular blue item is a lake or pond.

  c. Use logic and color to identify each map feature.
Note: For example, blue represents water. If you see a symbol that is blue and has clumps of grass, this would be a swamp.

 3. Identify the marginal information on a military map (Figure 071-COM-1000-2).

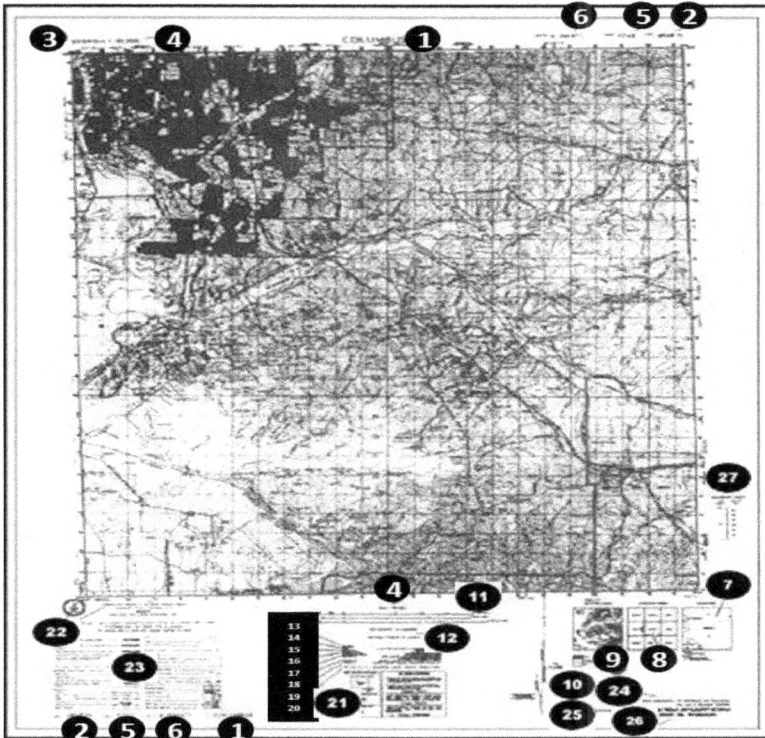

Figure 071-COM-1000-2. Topographical map.

  a. Identify the sheet name (1).
  b. Identify the sheet number (2).
  c. Identify the series name (3).
  d. Identify the scale (4).
  e. Identify the series number (5).
  f. Identify the edition number (6).
  g. Identify the index to boundaries (7).
  h. Identify the adjoining sheets diagram (8).
  i. Identify the elevation guide (9).
  j. Identify the declination diagram (10).
  k. Identify the bar scales (11).
  l. Identify the contour interval note (12).
  m. Identify the spheroid note (13).
  n. Identify the grid note (14).
  o. Identify the projection note (15).
  p. Identify the vertical datum note (16).
  q. Identify the horizontal datum note (17).
  r. Identify the control note (18).
  s. Identify the preparation note (19).
  t. Identify the printing note (20).
  u. Identify the grid reference box (21).
  v. Identify the unit imprint and symbol (22).
  w. Identify the legend (23).

Evaluation Preparation:

Setup: Provide the Soldier with the equipment and or materials described in the conditions statement.

Brief Soldier: Tell the Soldier what is expected of him by reviewing the task standards. Stress to the Soldier the importance of observing all cautions, warnings, and dangers to avoid injury to personnel and, if applicable, damage to equipment.

Performance Measures	**GO**	**NO GO**
1 Identified the six basic colors on a military map.	____	____
2 Identified the symbols on a military map.	____	____
3 Identified the marginal information on a military map.	____	____

Evaluation Guidance: Score the Soldier GO if all performance measures are passed. Score the Soldier NO-GO if any performance measure is failed. If the Soldier scores a NO-GO, show the Soldier what was done wrong and how to do it correctly.

References:
Required:
Related: TC 3-25.26

071-COM-1001

Identify Terrain Features on a Map

Foreign Disclosure: FD7 - This product/publication has been reviewed by the product developers in coordination with the DOTD, MCoE, Ft Benning, GA 31905 foreign disclosure authority. This product is NOT releasable to students from foreign countries.

Conditions: You are a member of a squad or team in a field environment and have been directed to identify the terrain features on a map. You have been given a 1:50,000 scale military map.

Standards: Identify the five major, three minor, and two supplementary terrain features on a military map.

Special Condition: None

Safety Risk: Low

MOPP 4:

Task Statements

Cue: None

*Note:*All terrain features are derived from a complex landmass known as a mountain or ridgeline (Figure 071-COM-1001-1). The term ridgeline is not interchangeable with the term ridge. A ridgeline is a line of high ground, usually with changes in elevation along its top and low ground on all sides from which a total of 10 natural or man-made terrain features are classified.

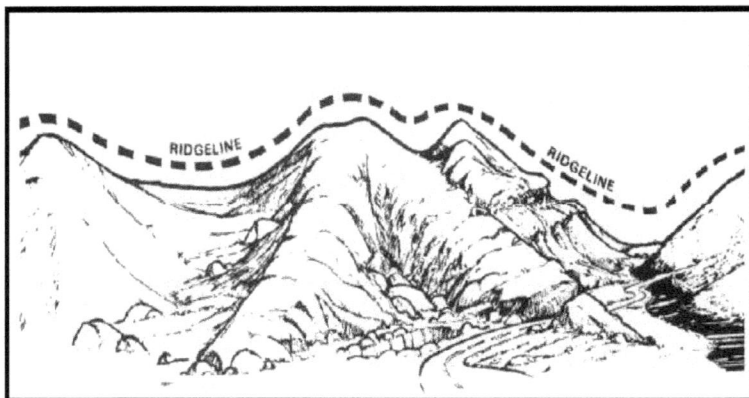

Figure 071-COM-1001-1. Ridgeline.

1. Identify five major terrain features.
   a. Identify a hill (Figure 071-COM-1001-2).
Note: A hill is an area of high ground. From a hilltop, the ground slopes down in all directions. A hill is shown on a map by contour lines forming concentric circles. The inside of the smallest closed circle is the hilltop.

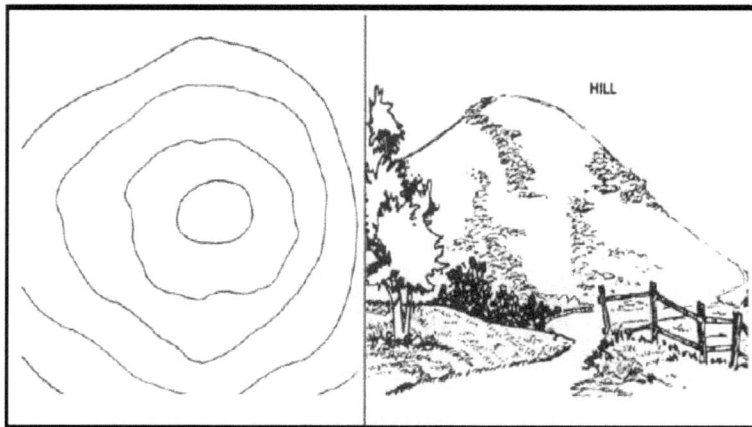

Figure 071-COM-1001-2. Hill.

   b. Identify a saddle (Figure 071-COM-1001-3).
Note: A saddle is a dip or low point between two areas of higher ground. A saddle is not necessarily the lower ground between two hilltops; it may be

simply a dip or break along a level ridge crest. If you are in a saddle, there is high ground in two opposite directions and lower ground in the other two directions. A saddle is normally represented as an hourglass.

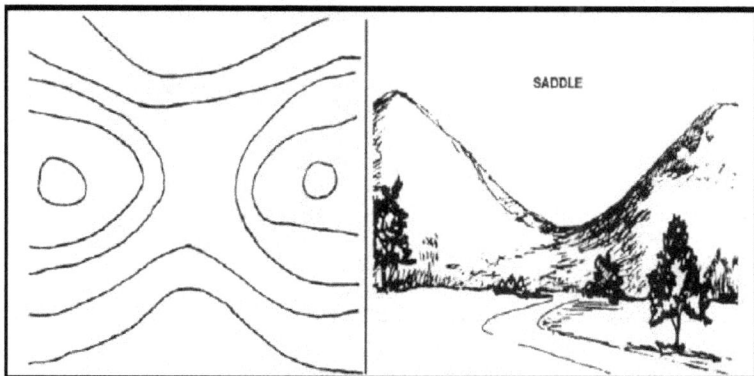

Figure 071-COM-1001-3. Saddle.

c. Identify a valley (Figure 071-COM-1001-4).

Note: A valley is a stretched-out groove in the land, usually formed by streams or rivers. A valley begins with high ground on three sides and usually has a course of running water through it. If standing in a valley, three directions offer high ground, while the fourth direction offers low ground. Depending on its size and where a person is standing, it may not be obvious that there is high ground in the third direction, but water flows from higher to lower ground. Contour lines forming a valley are either U-shaped or V-shaped. To determine the direction water is flowing, look at the contour lines. The closed end of the contour line (U or V) always points upstream or toward high ground.

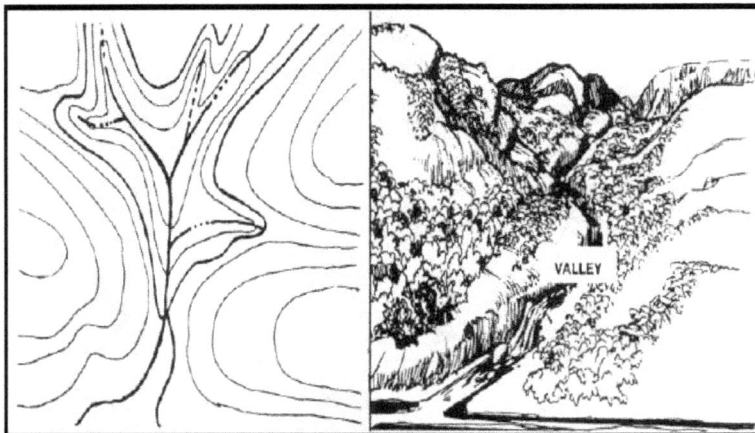

Figure 071-COM-1001-4. Valley.

d. Identify a ridge (Figure 071-COM-1001-5).
Note: A ridge is a sloping line of high ground. If you are standing on the centerline of a ridge, you will normally have low ground in three directions and high ground in one direction with varying degrees of slope. If you cross a ridge at right angles, you will climb steeply to the crest and then descend steeply to the base. When you move along the path of the ridge, depending on the geographic location, there may be either an almost unnoticeable slope or a very obvious incline. Contour lines forming a ridge tend to be U-shaped or V-shaped. The closed end of the contour line points away from high ground.

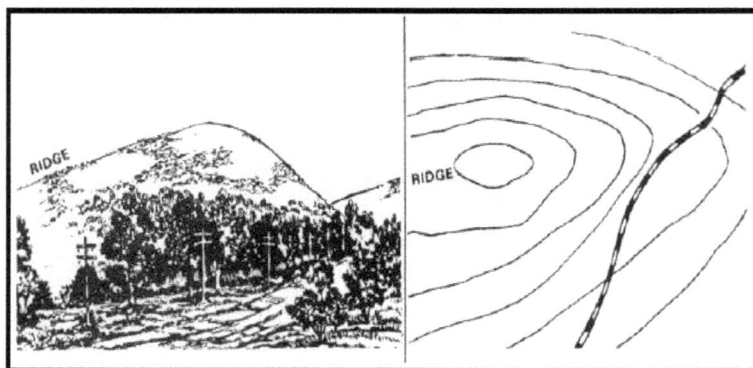

Figure 071-COM-1001-5. Ridge.

 e. Identify a depression (Figure 071-COM-1001-6).

Note: A depression is a low point in the ground or a sinkhole. It could be described as an area of low ground surrounded by higher ground in all directions, or simply a hole in the ground. Usually only depressions that are equal to or greater than the contour interval will be shown. On maps, depressions are represented by closed contour lines that have tick marks pointing toward low ground.

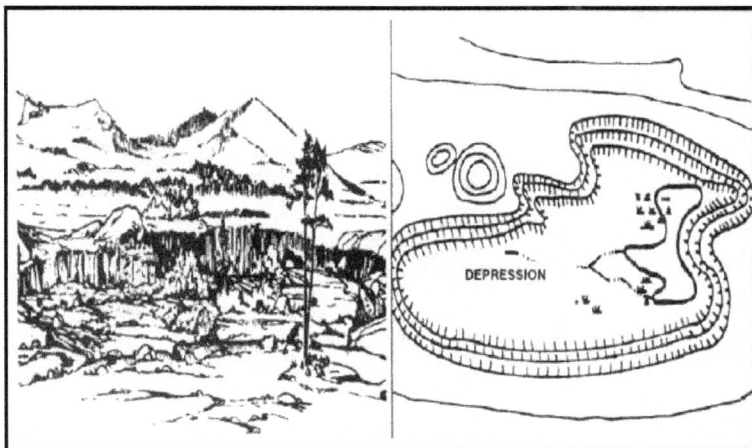

Figure 071-COM-1001-6. Depression.

 2. Identify three minor terrain features.

 a. Identify a draw (Figure 071-COM-1001-7).

Note: A draw is a stream course that is less developed than a valley. In a draw, there is essentially no level ground and, therefore, little or no maneuver room within its confines. If you are standing in a draw, the ground slopes upward in three directions and downward in the other direction. A draw could be considered as the initial formation of a valley. The contour lines depicting a draw are U-shaped or V-shaped, pointing toward high ground.

Figure 071-COM-1001-7. Draw.

b. Identify a spur (Figure 071-COM-1001-8).
Note: A spur is a short, continuous sloping line of higher ground, normally jutting out from the side of a ridge. A spur is often formed by two roughly parallel streams cutting draws down the side of a ridge. The ground will slope down in three directions and up in one. Contour lines on a map depict a spur with the U or V pointing away from high ground.

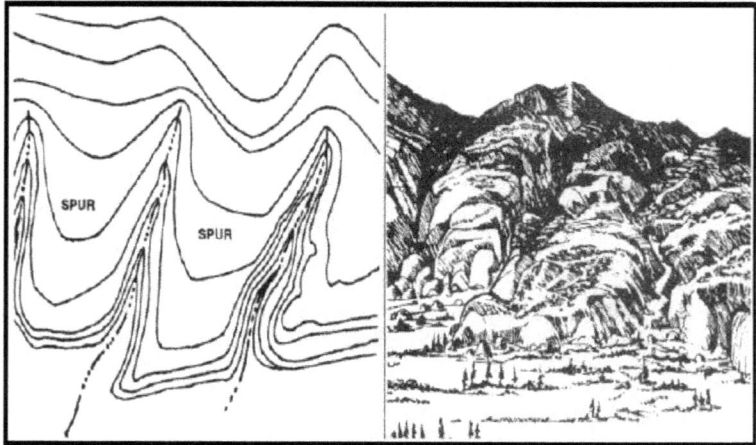

Figure 071-COM-1001-8. Spur.

c. Identify a cliff (Figure 071-COM-1001-9).
Note: A cliff is a vertical or near vertical feature; it is an abrupt change of the land. When a slope is so steep that the contour lines converge into one "carrying" contour of contours, this last contour line has tick marks pointing toward low ground. Cliffs re also shown by contour lines very close together

and, in some instances, touching each other.

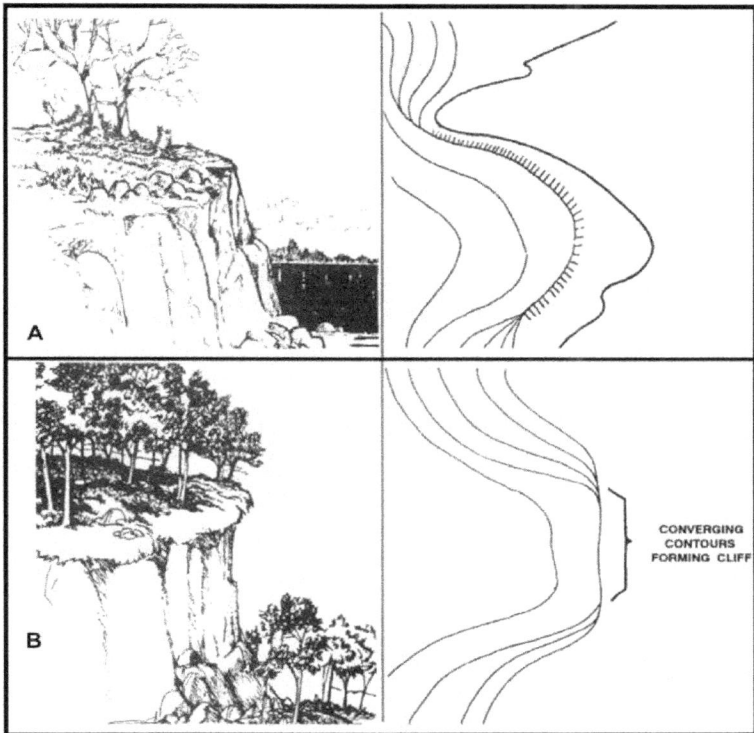

Figure 071-COM-1001-9. Cliff.

3. Identify two supplementary terrain features.

a. Identify a cut (Figure 071-COM-1001-10).

Note: A cut is a man-made feature resulting from cutting through raised ground, usually to form a level bed for a road or railroad track. Cuts are shown on a map when they are at least 10 feet high, and they are drawn with a contour line along the cut line. This contour line extends the length of the cut and has tick marks that extend from the cut line to the roadbed, if the map scale permits this level of detail.

b. Identify a fill (Figure 071-COM-1001-10).

Note: A fill is a man-made feature resulting from filling a low area, usually to form a level bed for a road or railroad track. Fills are shown on a map when they are at least 10 feet high, and they are drawn with a contour line along the fill line. This contour line extends the length of the filled area and has tick marks that point toward lower ground. If the map scale permits , the length of the fill tick marks are drawn to scale and extend from the base line of the fill symbol.

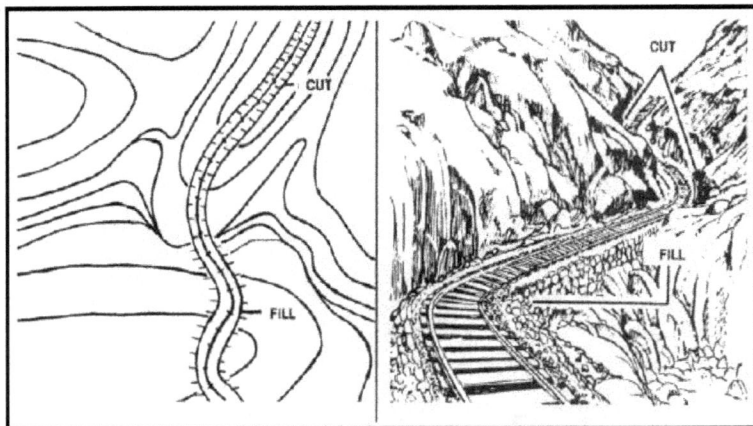

Figure 071-COM-1001-10. Cut and Fill.

Evaluation Preparation:

Setup: Provide the Soldier with the equipment and or materials described in the conditions statement.

Brief Soldier: Tell the Soldier what is expected of him by reviewing the task standards. Stress to the Soldier the importance of observing all cautions, warnings, and dangers to avoid injury to personnel and, if applicable, damage to equipment.

Performance Measures	GO	NO GO
1 Identified the five major terrain features.	_____	_____
2 Identified the three minor terrain features.	_____	_____
3 Identified the two supplementary terrain features.	_____	_____

Evaluation Guidance: Score the Soldier GO if all performance measures are passed. Score the Soldier NO-GO if any performance measure is failed. If the

Soldier scores a NO-GO, show the Soldier what was done wrong and how to do it correctly. n

References:
Required:
Related: TC 3-25.26

071-COM-1002

Determine the Grid Coordinates of a Point on a Military Map

Foreign Disclosure: FD7 - This product/publication has been reviewed by the product developers in coordination with the DOTD, MCoE, Ft Benning, GA 31905 foreign disclosure authority. This product is NOT releasable to students from foreign countries.

Conditions: You are a member of a squad or team in a field environment and have been directed to identify the grid coordinates of a point on a map. You have a 1:50,000 scale military map, a coordinate scale and protractor or plotting scale, a pencil, and paper. You have been shown the point on the map. Some iterations of this task should be performed in MOPP 4.

Standards: Determine the coordinates of the grid square, determine grid coordinates of a point with and without a coordinate scale and protractor or plotting scale. Identify the 100,000 meter square identifier to determine grid coordinate.

Special Condition: None

Special Standards: None

Safety Risk: Low

MOPP 4: Sometimes

Task Statements

Cue: None

Note: None

1. Determine the coordinates of the grid square (Figure 071-COM-1002-1).

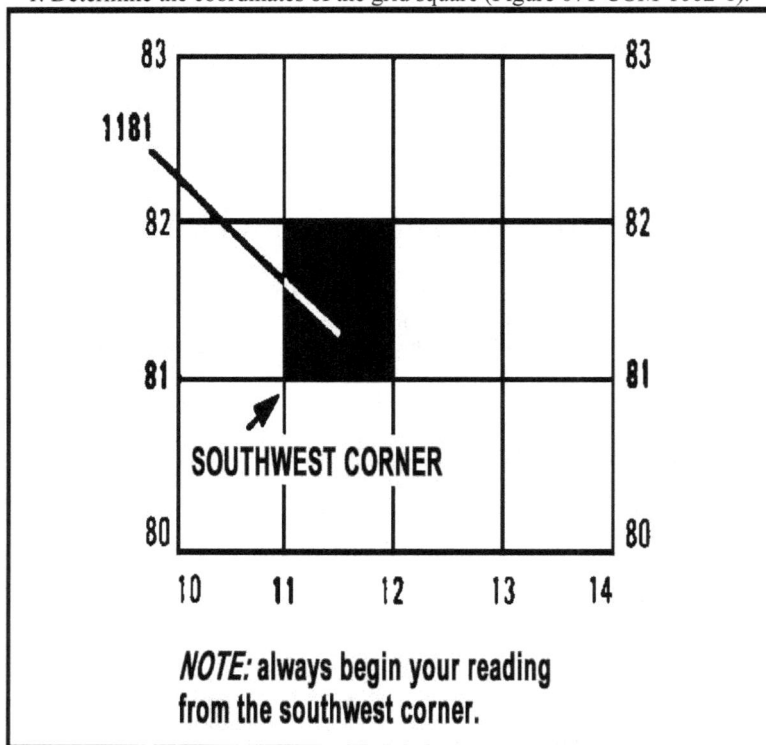

Figure 071-COM-1002-1. Identifying the Grid Square

a. Select the grid square that contains the identified point on the map (see Figure 071-COM-1002-1).

b. Read the north-south grid line that precedes the desired point (see Figure 071-COM-1002-1).

c. Record the number associated with that line.

d. Read the east-west grid line that precedes the desired point (see Figure 071-COM-1002- 1).

e. Record the number associated with that line.

Note: The number of digits represents the degree of precision to which a point has been located and measured on a map the more digits the more precise the measurement. In the above example the four digits 1181 identify the 1,000 meter grid square to be used.

2. Determine point grid coordinates without a coordinate scale and protractor or plotting scale (Figure 071-COM-1002-2).

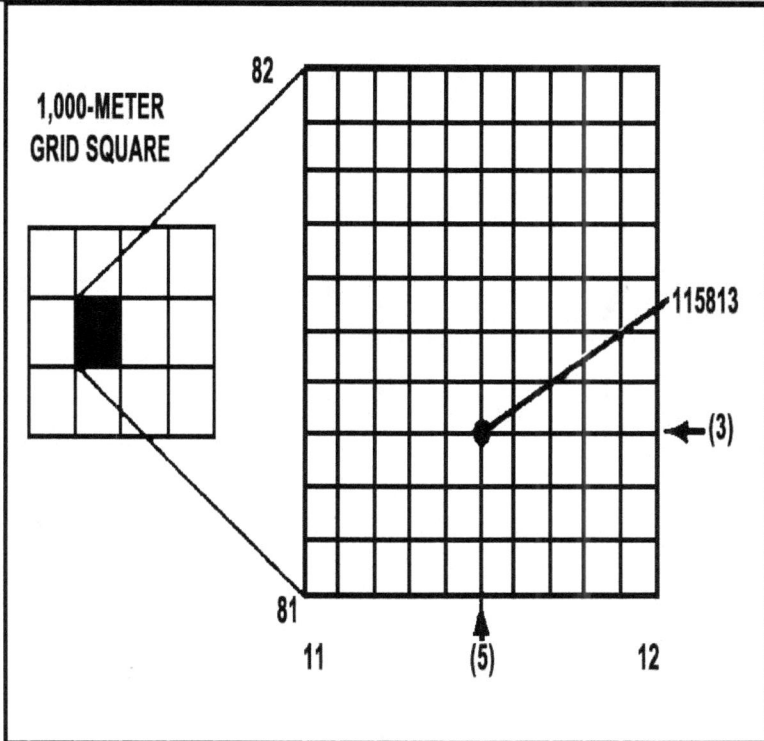

Figure 071-COM-1002-2. Grid Square 1181 Divided.

a. Allocate the grid square into a 10 by 10 grid.

b. Read right (from the lower left corner) to the imaginary gird line nearest the identified point.

Note: In the example the North-South imaginary line nearest the point is halfway or 5 lines out of a total of 10 lines. Therefore the first half of your grid coordinate is 115.

c. Read up (from the point reached in step 3b) to the imaginary gird line nearest the identified point.

Note: In the example the East-West imaginary line nearest the point is one third of the way up or 3 lines out of 10 lines. Therefore the second half of your grid coordinate is 813.

3. Determine point grid coordinates with GTA 05-02-012 coordinate scale and protractor or plotting scale (Figure 071-COM-1002-3).

Note: The most accurate way to determine the coordinates of a point on a map is with a coordinate scale. You need not imagine lines, because you can find the exact coordinates using the coordinate scale, protractor or the plotting scale.

Each device actually includes two coordinate scales, 1:25,000 and 1:50,000 meters. Make sure that, regardless which device you use, you choose the correct scale.

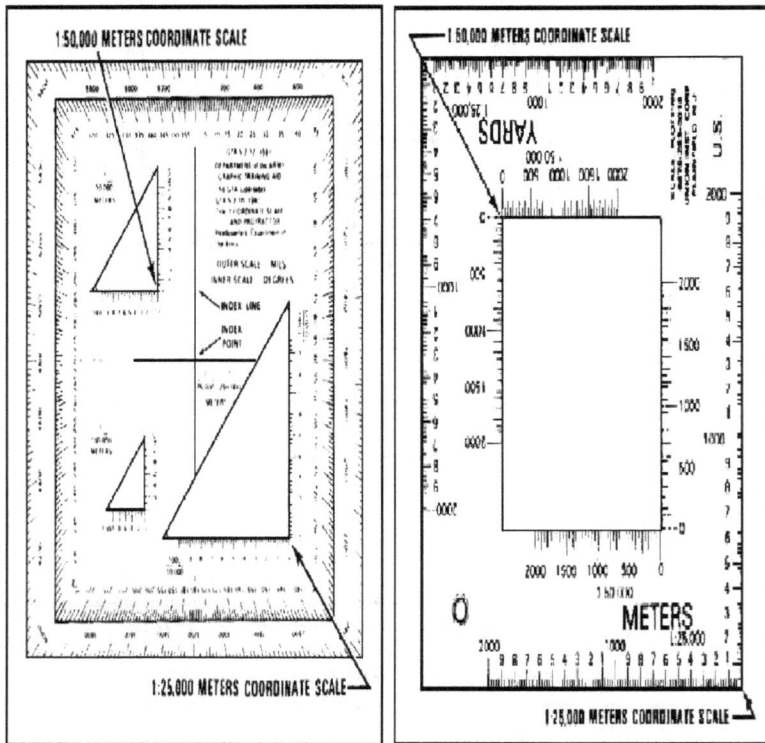

Figure 071-COM-1002-3. GTA 05-02-012 Coordinate Scale and Protractor (Left) and Plotting scale (Right).

    a. Locate the grid square where the point is located (Example: Point A in Figure 071-COM-1002-4).

    b. Determine the coordinates of the grid square.

Note: The number of the vertical grid line on the left (west) side of the grid square gives the first and second digits of the coordinate. The number of the horizontal grid line on the bottom (south) side of the grid square gives the fourth and fifth digits of the coordinate.

    c. Determine the third and sixth digits of the coordinate.

       (1) Place a coordinate scale and protractor or a plotting scale on the bottom horizontal grid line of the grid square containing Point A.

(2) Check to see that the zeros of the coordinate scale are in the lower left-hand (southwest) corner of the grid square where Point A is located (Figure 071-COM-1002-4).

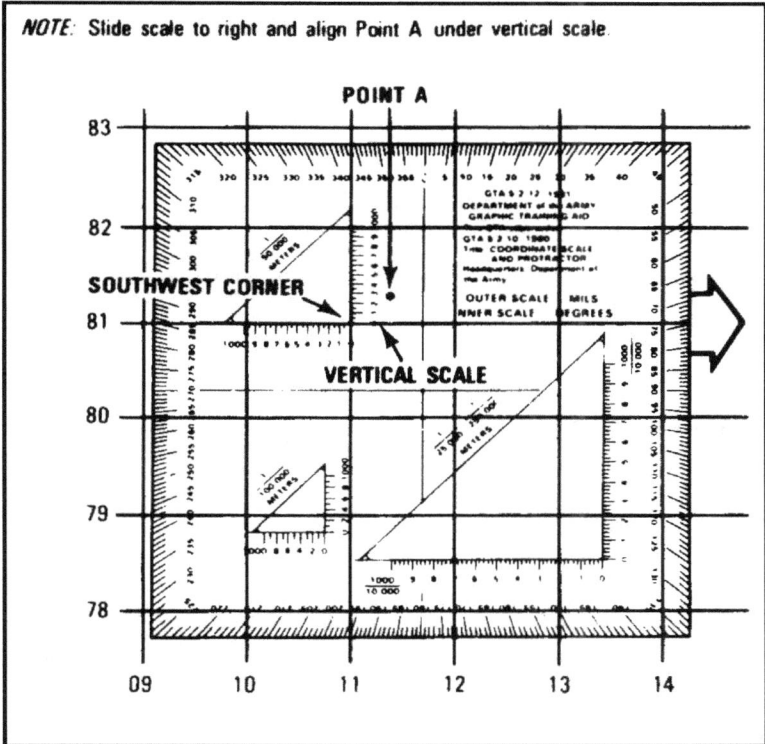

Figure 071-COM-1002-4. Placement of the Coordinate Scale.

(3) Slide the scale to the right, keeping the bottom of the scale on the bottom grid line until Point A is under the vertical (right-hand) scale (Figures 071-COM-1002-5 and 071-COM-1002-6).

Note: To determine the six-digit coordinate, look at the 100-meter mark on the bottom scale, which is nearest the vertical grid line. This mark is the third digit of the number 115. The 100-meter mark on the vertical scale nearest to Point A gives you the sixth digit of the number 813. The complete grid coordinate is 115813. Always read right, and then up.

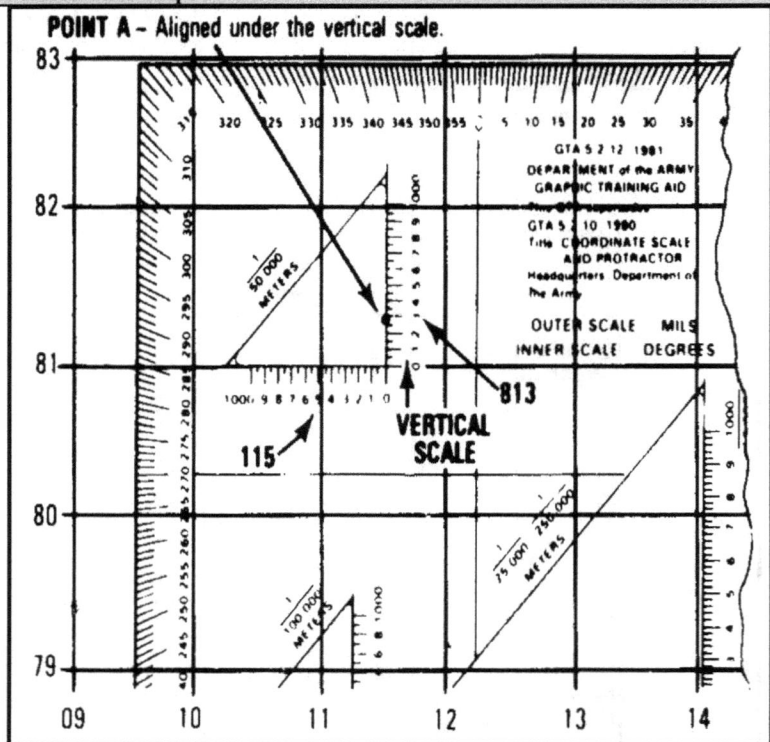

Figure 071-COM-1002-5. Aligning the Coordinate Scale.

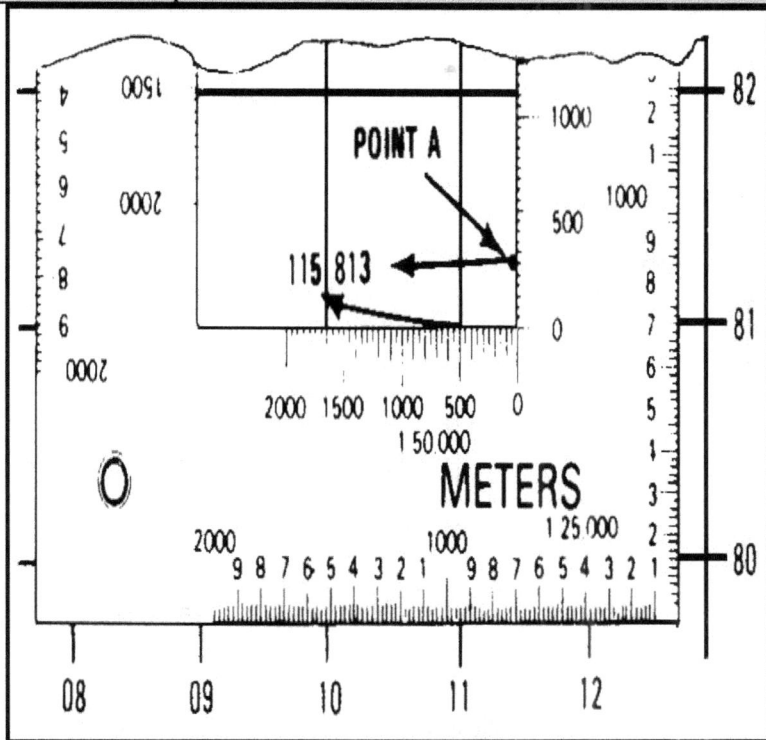

Figure 071-COM-1002-6. Aligning the Plotting Scale.

4. Add the two letter 100,000 meter square identifier to determined grid coordinate.

a. Identify the two letter 100,000 meter square identifier by looking at the grid reference box in the margin of the map (Figure 071-COM-1002-7).

SAMPLE 1,000-METER GRID SQUARE

46

X Sample
point

45

12   13

100,000-METER SQUARE IDENTIFICATION

FL | GL
^{7}00

GRID ZONE DESIGNATION

16S

100-METER REFERENCE

1. Read large numbers labeling the VERTICAL grid line left of point and estimate tenths (100-meters) from grid line to point.

2. Read large numbers labeling the HORIZONTAL grid line below point and estimate (100-meters) from grid line.

Example: 123456

WHEN REPORTING ACROSS A 100,000-METER LINE, PREFIX THE 100,000-METER SQUARE IDENTIFICATION, IN WHICH THE POINT LIES.

Example: FL123456

WHEN REPORTING OUTSIDE THE GRID ZONE DESIGNATION AREA, PREFIX THE GRID ZONE DESIGNATION.

Example: 16SFL123456

Figure 071-COM-1002-7. Grid Reference Box.

b. Place the 100,000 meter square identifier in front of the grid coordinate. *Note*: In the example given the final grid coordinate becomes GL115813.

Evaluation Preparation:

Setup: Provide the Soldier with the equipment and or materials described in the conditions statement.

Brief Soldier: Tell the Soldier what is expected of him by reviewing the task standards. Stress to the Soldier the importance of observing all cautions, warnings, and dangers to avoid injury to personnel and, if applicable, damage to equipment.

Performance Measures	GO	NO GO
1 Determined the coordinates of the grid square.	____	____
2 Determined point grid coordinates without a coordinate scale and protractor or plotting scale.	____	____
3 Determined point grid coordinates with coordinate scale and protractor or plotting scale.	____	____

Performance Measures	GO	NO GO
4 Added the two letter 100,000 meter square identifier to the determined grid coordinate.	_____	_____

Evaluation Guidance: Score the Soldier GO if all performance measures are passed. Score the Soldier NO-GO if any performance measure is failed. If the Soldier scores a NO-GO, show the Soldier what was done wrong and how to do it correctly.

References:
Required:
Related: TC 3-25.26

071-COM-1008

Measure Distance on a Map

Foreign Disclosure: FD7 - This product/publication has been reviewed by the product developers in coordination with the DOTD, MCoE, Ft Benning, GA 31905 foreign disclosure authority. This product is NOT releasable to students from foreign countries.

Conditions: You are a member of a squad or team in a field environment and have been directed to determine the distance between two known points. You have a 1:50,000 scale map, a strip of paper with a straight edge, and a pencil. You have been shown the beginning and ending points on the map. Some iterations of this task should be performed in MOPP 4.

Standards: Determine the straight-line distance between two points with no more than a 5 percent error and the road (curved line) distance between two points with no more than a 10 percent error.

Special Condition: None

Special Standards: None

Safety Risk: Low

MOPP 4: Sometimes

Task Statements

Cue: None

Note: None

Performance Steps

Note: The graphic scale is a ruler printed on the map and is used to convert distances on the map to actual ground distances. The graphic scale is divided into two parts. To the right of the zero, the scale is marked in full units of measure and is called the primary scale. To the left of the zero, the scale is divided into tenths and is called the extension scale. Most maps have three or more graphic scales, each using a different unit of measure. When using the graphic scale be sure to use the correct scale for the unit of measure desired.1. Identify the graphic (bar) scale of the map.

2. Determine straight-line distance between two points on a map.

    a. Line up the straight edge of a strip of paper with the beginning and ending points on the map.

    b. Mark a tick mark on the beginning and ending points on the straight edge of the paper (Figure 071-COM-1008-1).

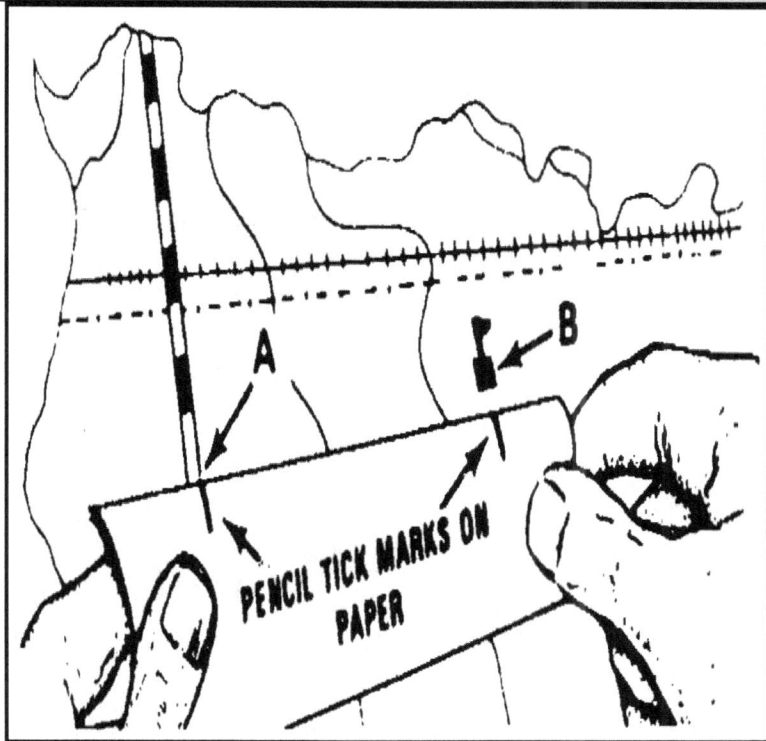

Figure 071-COM-1008-1. Beginning and Ending Points.

    c. Place the starting point on the paper under the zero on the bar scale.

    d. Measure off 4,000 meters and place a new tick mark on the paper.

    e. Place the new tick mark under the zero on the bar scale.

    f. Determine if the end point falls within the bar scale.

        (1) Record the value on the scale of the end point, if the end point fits on the scale.

        (2) Add 4,000 meters to this value (a) to get the total difference.

    g. Determine if the end point falls outside the bar.

        (1) Repeat steps 3d and 3e until the end point falls within the bar.

        (2) Add 4,000 meters to the value you derived in step 3f(1) for each time you performed step 3d to achieve the total distance.

    3. Convert map distance to ground distance.

    a. Align the edge of a strip of paper with the beginning point and the point where the road makes the first curve on the map.

    b. Mark on the straight edge of the paper the beginning and curve points.

    c. Repeat steps 4a and b, each time using the point of the curve as the next beginning point, until you reach the end point.

d. Align the marks on the paper with the appropriate bar scale (Figure 071-COM-1008-2).

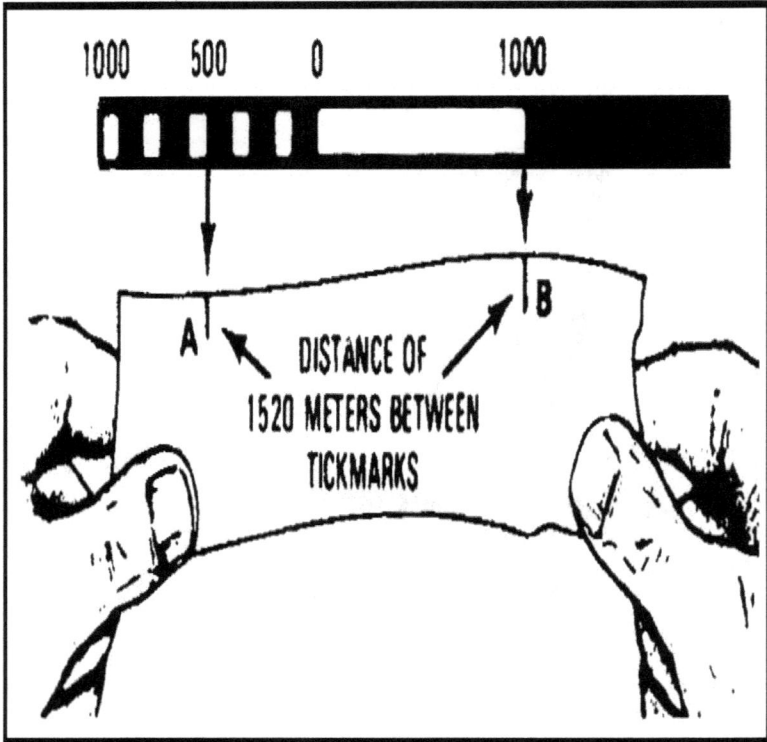

Figure 071-COM-1008-2. Distance between Beginning and Ending Points.

e. Determine the distance on the scale that compares to the distance on the paper.

4. Convert a road map distance to miles, meters or yards.

a. Align the edge of a strip of paper with the beginning point and the point where the road makes the first curve on the map.

b. Mark on the straight edge of the paper the beginning and curve points.

c. Repeat steps 5a and b, each time using the point of the curve as the next beginning point, until you reach the end point.

d. Place the starting point on the paper under the zero on the bar scale.

e. Measure off 4,000 meters and place a new tick mark on the paper.

f. Place the new tick mark under the zero on the bar scale.

g. Determine if the end point falls within the bar scale.

(1) Record the value on the scale of the end point, if the end point fits on the scale.

(2) Add 4,000 meters to this value (a) to get the total difference.

h. Determine if the end point falls outside the bar.

(1) Repeat steps 5d and 5e until the end point falls within the bar.

(2) Add 4,000 meters to the value you derived in step 5g(1) for each time you performed step 5d to achieve the total distance.

Evaluation Preparation:

Setup: Provide the Soldier with the equipment and or materials described in the conditions statement.

Brief Soldier: Tell the Soldier what is expected of him by reviewing the task standards. Stress to the Soldier the importance of observing all cautions, warnings, and dangers to avoid injury to personnel and, if applicable, damage to equipment.

Performance Measures	**GO**	**NO GO**
1 Identified the graphic bar scale of the map.	_____	_____
2 Determine the straight-line distance between two points on a map.	_____	_____
3 Measure the distance along a road, stream, or other curved line.	_____	_____

Evaluation Guidance: Score the Soldier GO if all performance measures are passed. Score the Soldier NO-GO if any performance measure is failed. If the Soldier scores a NO-GO, show the Soldier what was done wrong and how to do it correctly.

References:
Required:
Related: TC 3-25.26

071-COM-1005

Determine a Location on the Ground by Terrain Association

Conditions: You are a member of a squad or team in a field environment and have been directed to determine your squad's/team's current location. You have a

1:50,000 scale military map, a compass, a coordinate scale and protractor or plotting scale, a pencil, and paper. Some iterations of this task should be performed in MOPP 4.

Standards: Orient the map. Identify the type of terrain on which you are located as well as the type of terrain that surrounds your location. Correlate the terrain features on the ground to those shown on the map. Determine the six digit grid coordinates to your location..

Special Condition: None

Special Standards: None

Safety Risk: Low

MOPP 4: Sometimes

Task Statements

Cue: None

Note: None

Performance Steps

    1. Orient the map.
Note: There are three ways to orient a map:
- Using a compass. The magnetic arrow of the compass points to magnetic north. As such, pay special attention to the declination diagram.
- Using terrain association. This method is typically used when a compass is not available or when the user has to make many quick references as he moves across country.
- Using Field-Expedient Methods. These methods are used when a compass is available and there are no recognizable terrain features.
    2. Identify the type of terrain feature on which you are located.
    3. Identify the types of terrain features that surround your location.
    4. Correlate the terrain features on the ground to those shown on the map.
    5. Determine your location on the map.
    6. Determine the six digit grid coordinate of your location.
Note: Grid coordinates of your location can be determined by using a coordinate scale and protractor, a plotting scale, or by visualizing a 10 by 10 grid box inside the appropriate grid square.

Evaluation Preparation:

Setup: Provide the Soldier with the equipment and or materials described in the conditions statement.

Brief Soldier: Tell the Soldier what is expected of him by reviewing the task standards. Stress to the Soldier the importance of observing all cautions, warnings, and dangers to avoid injury to personnel and, if applicable, damage to equipment.

Performance Measures	GO	NO GO
1 Oriented the map.	_____	_____
2 Identified the type of terrain feature on which you were located.	_____	_____
3 Identified the types of terrain features that surround your location.	_____	_____
4 Correlated the terrain features on the ground to those shown on the map.	_____	_____
5 Determined your location on the map.	_____	_____
6 Determined the six digit grid coordinate of your location.	_____	_____

Evaluation Guidance: Score the Soldier GO if all performance measures are passed. Score the Soldier NO-GO if any performance measure is failed. If the Soldier scores a NO-GO, show the Soldier what was done wrong and how to do it correctly.

References
Required:
Related: TC 3-25.26

071-COM-1012

Orient a Map to the Ground by Map-Terrain Association

Conditions: You are a member of a squad or team that is conducting movement in a field environment and you have been directed to orient a standard 1:50,000 scale military map to the ground. You do not have an operational compass. Some iterations of this task should be performed in MOPP 4.

Standards: Hold the map horizontally and match terrain features appearing on the map with physical features on the ground. Orient the map to within 30 degrees of magnetic north.

Special Condition: None

Special Standards: None

Safety Risk: Low

Task Statements

Cue: None

Note: A map can be oriented by terrain association when a compass is not available or when the user has to make many quick references as he moves across country. Using this method requires careful examination of the map and the ground, and the user must know his approximate location.

Performance Steps
1. Hold the map in a horizontal position.
2. Match terrain features appearing on your map with terrain features physically observable on the ground (Figure 071-COM-1012-1).

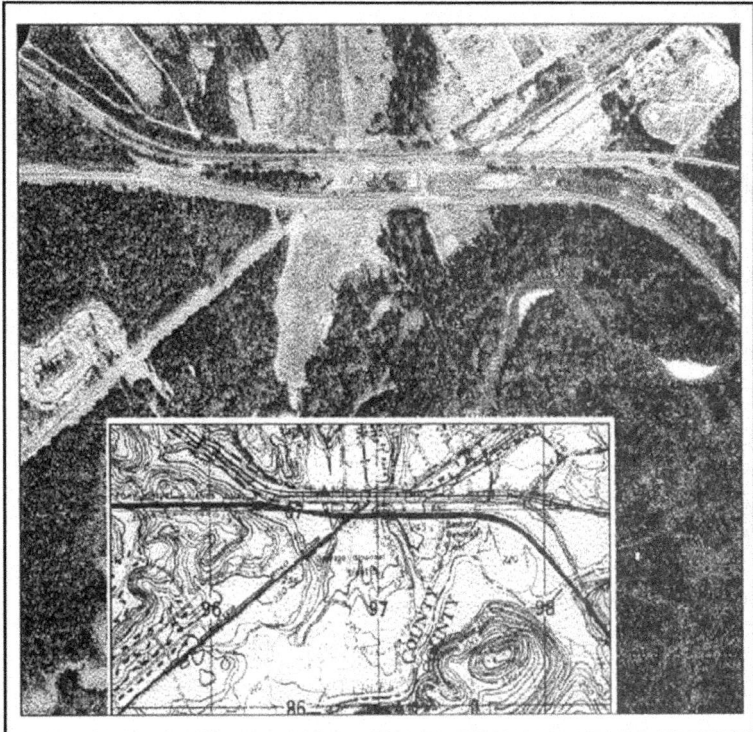

Figure 071-COM-1012-1. Terrain Association.

3. Align the map with the terrain features on the ground.

Evaluation Preparation:

Setup: Provide the Soldier with the equipment and or materials described in the conditions statement.

Brief Soldier: Tell the Soldier what is expected of him by reviewing the task standards. Stress to the Soldier the importance of observing all cautions, warnings, and dangers to avoid injury to personnel and, if applicable, damage to equipment.

Performance Measures	GO	NO GO
1 Held the map in a horizontal position.	_____	_____

Performance Measures	GO	NO GO
2 Matched terrain features appearing on map with physical features on the ground.	_____	_____
3 Aligned the map with the terrain features on the ground to within 30 degrees of magnetic north.	_____	_____

Evaluation Guidance: Score the Soldier GO if all performance measures are passed. Score the Soldier NO-GO if any performance measure is failed. If the Soldier scores a NO-GO, show the Soldier what was done wrong and how to do it correctly.

References:
Required:
Related: TC 3-25.26

071-COM-1011

Orient a Map Using a Lensatic Compass

Conditions: You are a member of a squad or team in a field environment and have been directed to orient a map in preparation for movement. You have a 1:50,000-scale topographic map of the area and a compass. Some iterations of this task should be performed in MOPP 4.

Standards: Determine the direction and value of declination, lay the map in a horizontal position and orient the map to the ground using a compass.

Special Condition: None

Special Standards: None

Safety Risk: Low

Task Statements

Cue: None

Note: The first step for a navigator in the field is orienting the map. A map is oriented when it is in a horizontal position with its north and south corresponding to the north and south on the ground.

When orienting a map with a compass, remember that the compass measures magnetic azimuths. Since the magnetic arrow points to magnetic north, pay special attention to the declination diagram. Two techniques are used.

Special care should be taken when orienting your map with a compass. A small mistake can cause you to navigate in the wrong direction.

Once the map is oriented, magnetic azimuths are determined using the compass. Do not move the map from its oriented position since any change in its position moves it out of line with the magnetic north.

Performance Steps

1. Determine the direction of the declination and its value from the declination diagram on the map.
2. Lay the map in a horizontal position.
3. Use one of the two techniques to orient the map.
    a. Orient the map using the first technique.
      (1) Take the straightedge on the left side of the compass and place it alongside the north-south grid line with the cover of the compass pointing toward the top of the map.
Note: This procedure places the fixed black index line of the compass parallel to north-south grid lines of the map.
      (2) Keep the compass aligned as directed above while rotating the map and compass together until the magnetic arrow is below the fixed black index line on the compass.
Note: At this time, the map is close to being oriented.
      (3) Rotate the map and compass in the direction of the declination diagram.
      (4) Verify the G-M angle.
        (a) If the magnetic north arrow on the map is to the left of the grid north, check the compass reading to see if it equals the G-M angle given in the declination diagram (Figure 071-COM-1011-1).

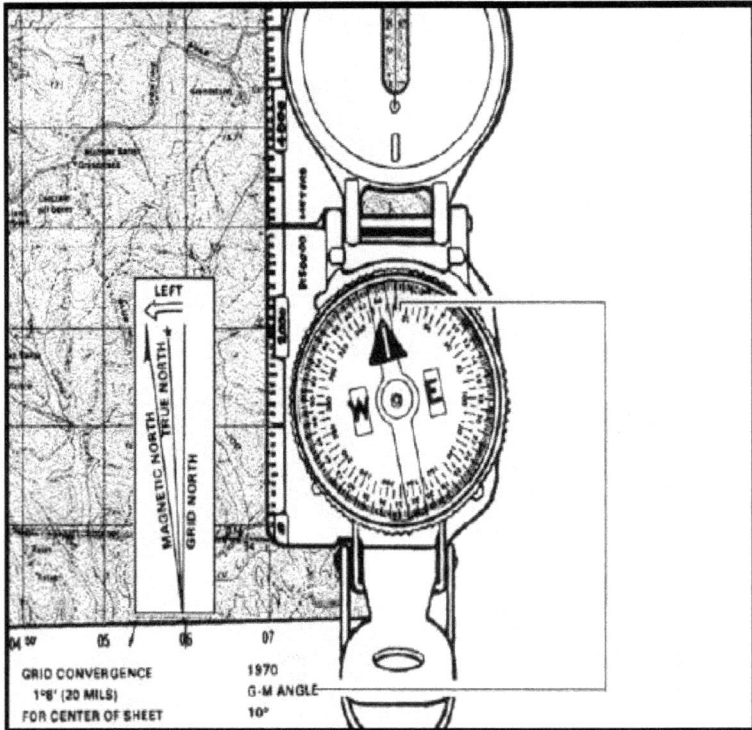

Figure 071-COM-1011-1. Map oriented with 10 degrees west declination.

(b) If the magnetic north is to the right of grid north, check the compass reading to see if it equals 360 degrees minus the G-M angle (Figure 071-COM-1011-2).

Note: If the G-M angles are correct the map is oriented.

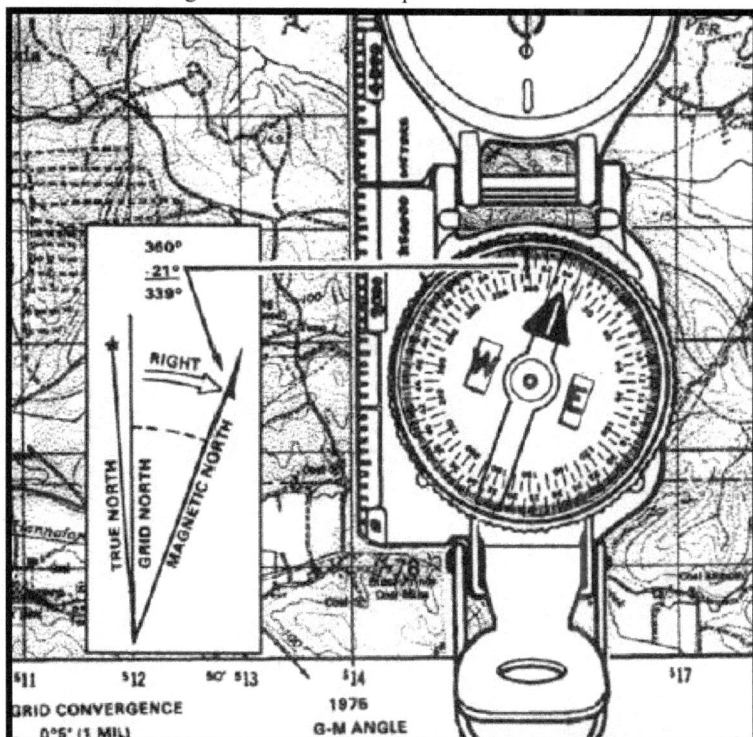

Figure 071-COM-1011-2. Map oriented with 21 degrees east declination.

b. Orient the map using the second technique.

(1) Draw a magnetic azimuth equal to the G-M angle given in the declination diagram with the protractor using any north-south grid line on the map as a base.

(2) If the declination is easterly (right), the drawn line is equal to the value of the G-M angle:

(a) Align the straightedge on the left side of the compass alongside the drawn line on the map.

(b) Rotate the map and compass until the magnetic arrow of the compass is below the fixed black index line (Figure 071-COM-1011-3).

Note: The map is now oriented.

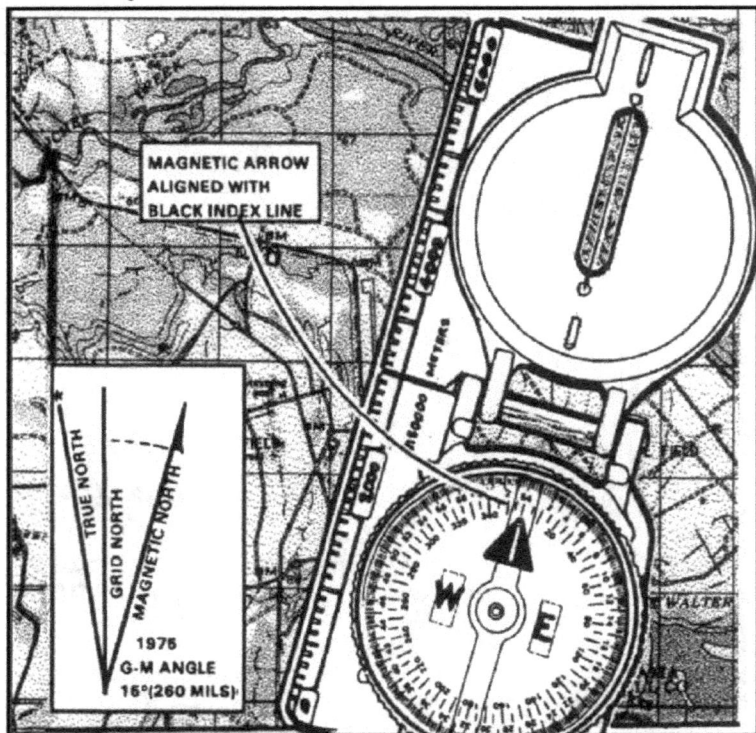

Figure 071-COM-1011-3. Map oriented with 15 degrees east declination.

      (3) If the declination is westerly (left), the drawn line will equal 360 degrees minus the value of the G-M angle:

      (a) Align the straightedge on the left side of the compass alongside the drawn line on the map.

      (b) Rotate the map and compass until the magnetic arrow of the compass is below the fixed black index line.

Note: The map is now oriented.

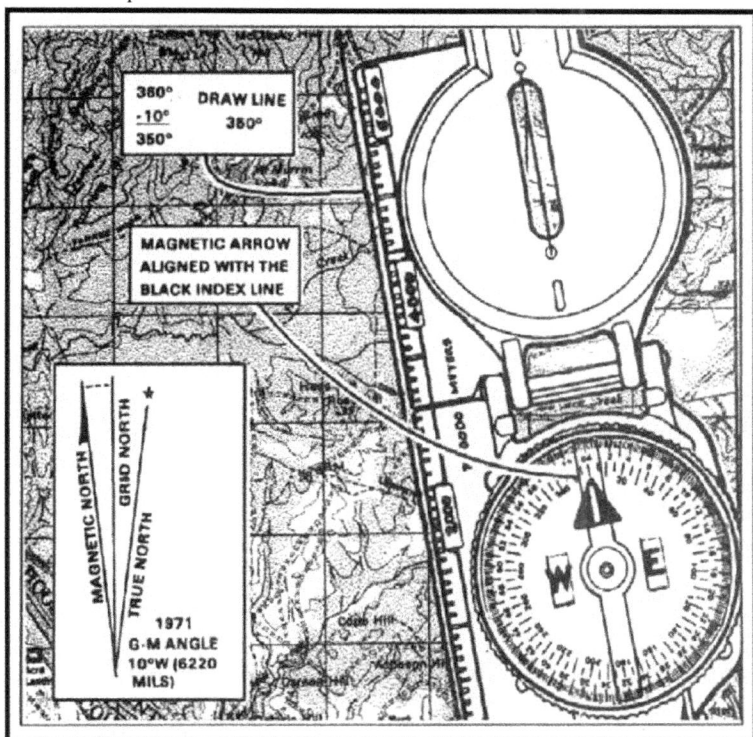

Figure 071-COM-1011-4. Map oriented with 10 degrees west declination.

Evaluation Preparation:

Setup: Provide the Soldier with the equipment and or materials described in the conditions statement.

Brief Soldier: Tell the Soldier what is expected of him by reviewing the task standards. Stress to the Soldier the importance of observing all cautions, warnings, and dangers to avoid injury to personnel and, if applicable, damage to equipment.

Performance Measures	GO	NO GO
1 Determined the direction of the declination and its value from the declination diagram.	_____	_____
2 Laid the map in a horizontal position.	_____	_____
3 Used one of the two techniques to orient the map.	_____	_____

Evaluation Guidance: Score the Soldier GO if all performance measures are passed. Score the Soldier NO-GO if any performance measure is failed. If the Soldier scores a NO-GO, show the Soldier what was done wrong and how to do it correctly.

References:
Required:
Related: TC 3-25.26

071-COM-1003

Determine a Magnetic Azimuth Using a Lensatic Compass

Conditions: You are a member of a squad or team in a field environment and have been directed to determine a magnetic azimuth. You have a compass and a designated point on the ground. Some iterations of this task should be performed in MOPP 4.

Standards: Inspect the compuss. Determine the correct magnetic azimuth to the designated point within 3 degrees using the compass-to-cheek method, and within 10 degrees using the center-hold method.

Special Condition: None

Special Standards: None

Safety Risk: Low

MOPP 4: Sometimes

Task Statements

Cue: None

Performance Steps

1. Inspect the compass (Figure 071-COM-1003-1).

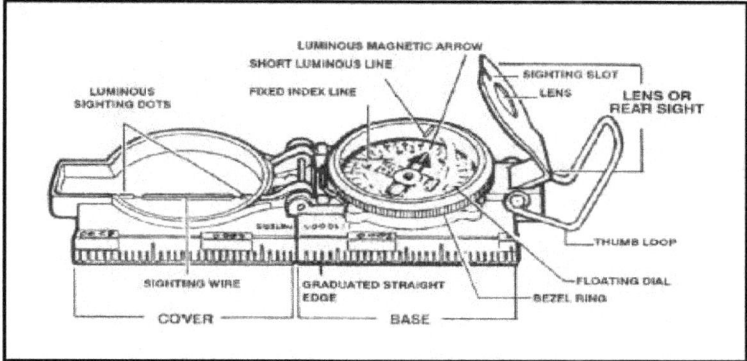

Figure 071-COM-1003-1. Lensatic compass.

    a. Ensure floating dial, which contains the magnetic needle moves freely and does not stick.

    b. Ensure the sighting wire is straight.

    c. Ensure glass and crystal parts are not broken.

    d. Ensure numbers on the dial are readable.

2. Determine direction (Figure 071-COM-1003-2).

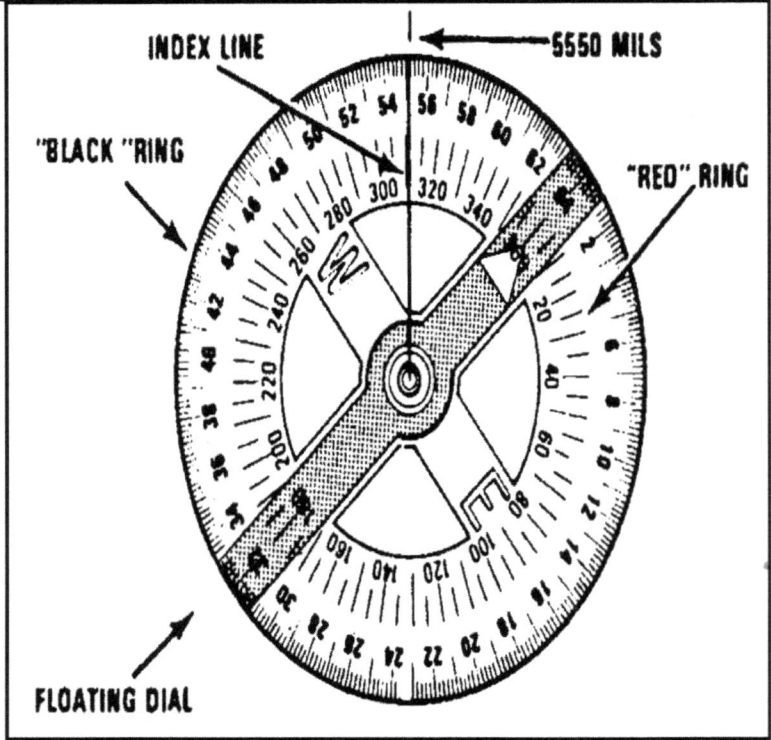

Figure 071-COM-1003-2. Lensatic compass floating dial.

a. Align the compass to the direction you want to go or want to determine.

b. Locate the scale beneath the index line on the outer glass cover.

c. Determine to the nearest degree, or 10 mils, the position of the index line over the red or black scale.

Note: Effects of Metal and Electricity. Metal objects and electrical sources can affect the performance of a compass. However, nonmagnetic metals and alloys do not affect compass readings. The following separation distances are suggested to ensure proper functioning of a compass:

High-tension power lines .... 55 meters.
Field gun, truck, or tank.. 18 meters.
Telegraph or telephone wires and barbed wire....... 10 meters.
Machine gun .... 2 meters.
Steel helmet or rifle... 1/2 meter.

3. Determine an azimuth with the compass-to-cheek method (Figure 071-COM-1003-3).

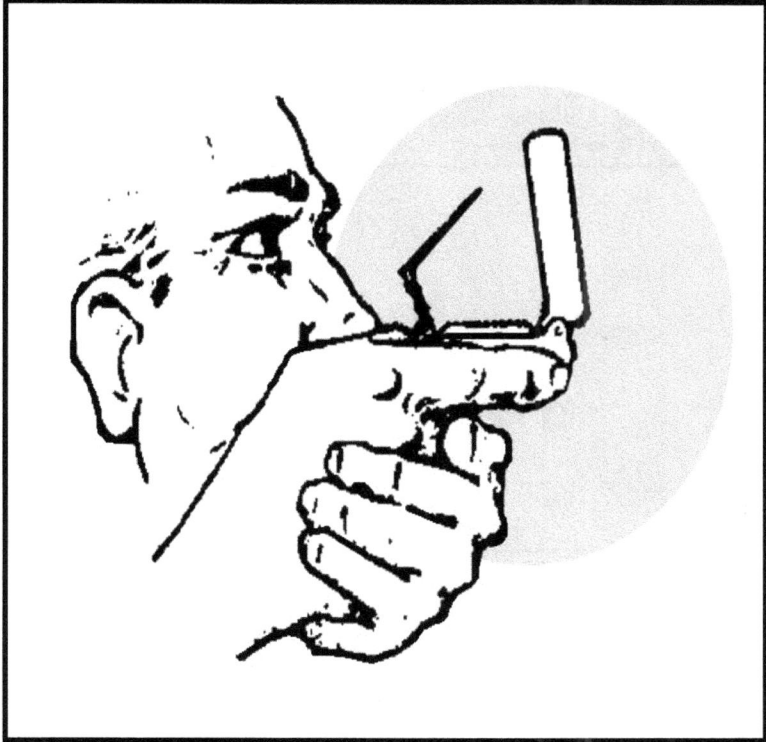

Figure 071-COM-1003-3. Compass-to-cheek method.

a. Open the cover to a 90-degree angle to the base.

b. Position the eyepiece at a 45-degree angle to the base.

c. Place your thumb through the thumb loop.

d. Establish a steady base with your third and fourth fingers.

e. Extend your index finger along the side of the compass base.

f. Place the hand holding the compass into the palm of the other hand.

g. Move both hands up to your face.

h. Position the thumb that is through the thumb loop against the cheekbone.

i. Move the eyepiece up or down until the dial is in focus.

j. Align the sighting slot of the eyepiece with the sighting wire in the cover on the desired point.

k. Read the azimuth under the index line.

4. Determine an azimuth with the center-hold method (Figure 071-COM-1003-4).

Note: This method offers the following advantages over the sighting technique:

-It is faster and easier to use.

-It can be used under all conditions of visibility.

-It can be used when navigating over any type of terrain.

-It can be used without putting down the rifle; however, the rifle must be slung well back over either shoulder.

-It can be used without removing eyeglasses

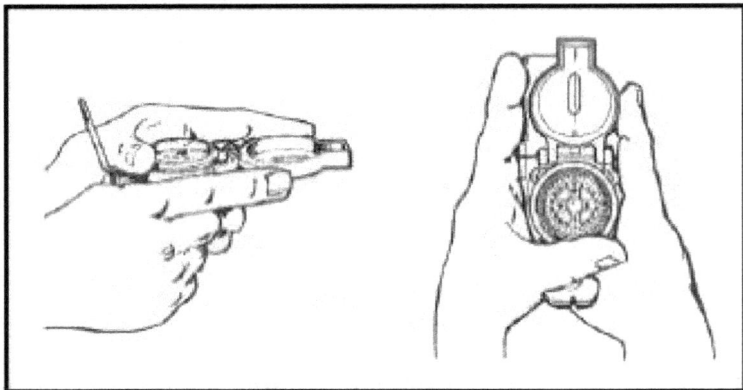

Figure 071-COM-1003-4. Centerhold technique.

a. Open the compass so that the cover forms a straight edge with the base.

b. Position the eyepiece lens to the full upright position.

c. Place your thumb through the loop.

d. Establish a steady base with your third and fourth fingers.

e. Extend your index finger along the side of the compass.

f. Place the thumb of your other hand between the eyepiece and lens.

g. Extend the index finger along the remaining side of the compass.

h. Secure the remaining fingers around the fingers of the other hand.

i. Place your elbows firmly into your side.

Note: This will place the compass between your chin and your belt.

j. Turn your entire body toward the object.

k. Align the compass cover directly at the object.

l. Read the azimuth from beneath the fixed black index line.

Evaluation Preparation:

Setup: Provide the Soldier with the equipment and or materials described in the conditions statement.

Brief Soldier: Tell the Soldier what is expected of him by reviewing the task standards. Stress to the Soldier the importance of observing all cautions, warnings, and dangers to avoid injury to personnel and, if applicable, damage to equipment.

Performance Measures	GO	NO GO
1 Inspected the compass.	_____	_____
2 Determined direction.	_____	_____
3 Determined an azimuth using the compass-to-cheek method.	_____	_____
4 Determined an azimuth using the center-hold method.	_____	_____

Evaluation Guidance: Score the Soldier GO if all performance measures are passed. Score the Soldier NO-GO if any performance measure is failed. If the Soldier scores a NO-GO, show the Soldier what was done wrong and how to do it correctly.

References:
Required:
Related: TC 3-25.26

071-COM-0018

Determine Grid Azimuth Using a Protractor

Conditions: You are a member of a squad or team that has received an order requiring movement. You have been directed to determine the grid azimuth from the starting point and ending point designated in the order. You have a 1:50,000-scale military map, a military protractor (GTA 05-02-012), a pencil, and paper. Some iterations of this task should be performed in MOPP 4.

Standards: Identify the starting point and ending point on the map, draw a straight line on the map connecting the points, align the protractor to the map, and determine the value of the angle.

Special Condition: None

Safety Risk: Low

| Task Statements |

Performance Steps

1. Identify the starting point (A) on the map.
2. Identify the ending point (B) on the map.

Note: When measuring azimuths on a map, remember that you are measuring from a starting point to an ending point. If a mistake is made and the reading is taken from the ending point, the grid azimuth will be 180 degrees off, thus causing the user to go in the wrong direction.

3. Draw a straight line on the map connecting the two points.

Note: To ensure an accurate measurement the line should extend past the starting point and ending point

4. Align the protractor to the map. (figure 074-COM-001-1 and 074-COM-001-2)

Note: When using the protractor, the base line is always oriented parallel to a north-south grid line. The 0- or 360-degree mark is always toward the top or north on the map and the 90-degree mark is to the right

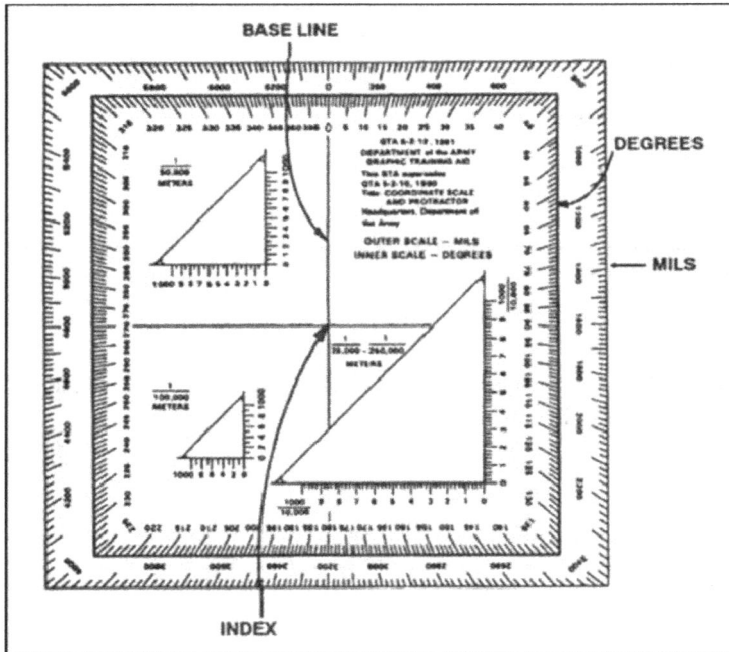

Figure 071-COM-0001-1. Protractor

1. Identify the starting point (A) on the map.

2. Identify the ending point (B) on the map.

Note: When measuring azimuths on a map, remember that you are measuring from a starting point to an ending point. If a mistake is made and the reading is taken from the ending point, the grid azimuth will be 180 degrees off, thus causing the user to go in the wrong direction.

3. Draw a straight line on the map connecting the two points.

Note: To ensure an accurate measurement the line should extend past the starting point and ending point

4. Align the protractor to the map. (figure 074-COM-001-1 and 074-COM-001-2)

Note: When using the protractor, the base line is always oriented parallel to a north-south grid line. The 0- or 360-degree mark is always toward the top or north on the map and the 90-degree mark is to the right

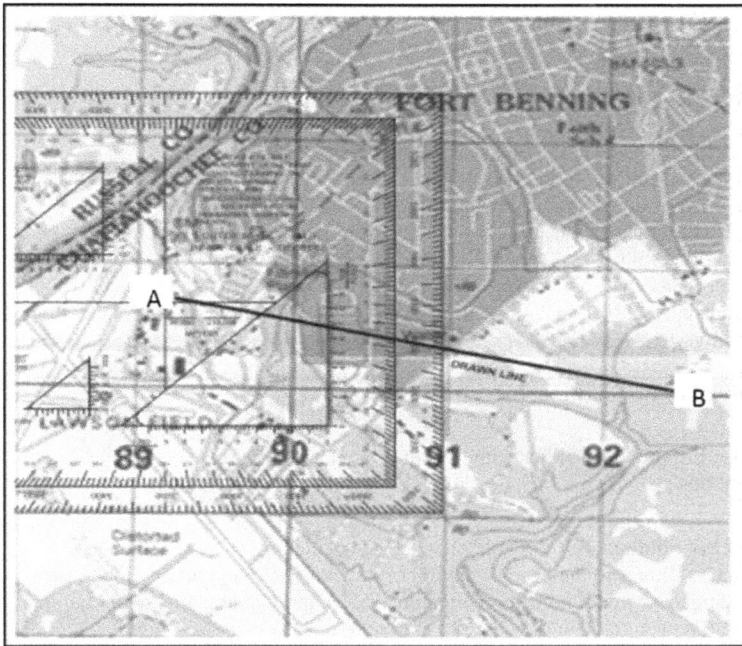

Figure 071-COM-0001-2. Measuring an azimuth.

a. Place the index of the protractor at the point where the drawn line crosses a vertical (north-south) grid line.

b. Align the 0- to 180-degree line of the protractor on the vertical grid line, while keeping the index in position.

Performance Steps

    1. Identify the starting point (A) on the map.

    2. Identify the ending point (B) on the map.

Note: When measuring azimuths on a map, remember that you are measuring from a starting point to an ending point. If a mistake is made and the reading is taken from the ending point, the grid azimuth will be 180 degrees off, thus causing the user to go in the wrong direction.

    3. Draw a straight line on the map connecting the two points.

Note: To ensure an accurate measurement the line should extend past the starting point and ending point

    4. Align the protractor to the map. (figure 074-COM-001-1 and 074-COM-001-2)

Note: When using the protractor, the base line is always oriented parallel to a north-south grid line. The 0- or 360-degree mark is always toward the top or north on the map and the 90-degree mark is to the right

    5. Determine the value of the angle (grid azimuth) from the scale.

Note: The grid azimuth is the degrees/mils value where the azimuth line crosses the protractor scale.

    a. Identify where the line crosses the protractor scale.

    b. Read the value (degrees or mils) where the line intersects with the scale.

Note: This value is the grid azimuth.

(Asterisks indicates a leader performance step.)

Evaluation Guidance: Score the Soldier GO if all performance measures are passed. Score the Soldier NO-GO if any performance measure is failed. If the Soldier scores a NO-GO, show the Soldier what was done wrong and how to do it correctly.

Evaluation Preparation:

Setup: Provide the Soldier with the equipment and/or materials described in the conditions statement.

Brief the Soldier: Tell the Soldier what is expected by reviewing the task standards. Stress to the Soldier the importance of observing all cautions, warnings, and dangers to avoid injury to personnel and, if applicable, damage to equipment.

Performance Measures	GO	NO GO
1 Identified the starting point on the map.	_____	_____

Performance Measures	GO	NO GO
2 Identify the ending point on the map.	_____	_____
3 Drew a straight line on the map connecting the two points.	_____	_____
4 Aligned the protractor to the map.	_____	_____
5 Determined the value of the angle.	_____	_____

Environment: Environmental protection is not just the law but the right thing to do. It is a continual process and starts with deliberate planning. Always be alert to ways to protect our environment during training and missions. In doing so, you will contribute to the sustainment of our training resources while protecting people and the environment from harmful effects. Refer to the current ATP 3-34.5 Environmental Considerations Manual and the current GTA 05-08-002 Environmental-Related Risk Assessment card.

Safety: In a training environment, leaders must perform a risk assessment in accordance with current Risk Management Doctrine. Leaders will complete the current Deliberate Risk Assessment Worksheet in accordance with the TRADOC Safety Officer during the planning and completion of each task and sub-task by assessing mission, enemy, terrain and weather, troops and support available-time available and civil considerations, (METT-TC). *Note:* During MOPP training, leaders must ensure personnel are monitored for potential heat injury. Local policies and procedures must be followed during times of increased heat category in order to avoid heat related injury. Consider the MOPP work/rest cycles and water replacement guidelines IAW current CBRN doctrine.

071-COM-1016

Convert Azimuths

Conditions: You are a member of a squad preparing to conduct tactical movement and you have been directed to convert azimuths in preparation for the movement. You have been given a 1:50,000 military map with a declination diagram, pencil, paper, a magnetic azimuth, and a grid azimuth. Some iterations of this task should be performed in MOPP 4.

Standards: Convert an azimuth into a back azimuth, a magnetic azimuth to a grid azimuth and a grid azimuth to a magnetic azimuth, without error.

Special Condition: None

Special Standards: None

Safety Risk: Low

MOPP 4: Sometimes

Task Statements

Cue: *None*

Note: The North-South lines designate grid North on your map. The compass needle points to magnetic North. The grid magnetic (G-M) angle is the angle difference between grid North and magnetic North.

Performance Steps

WARNING

When converting azimuths into back azimuths, extreme care should be exercised when adding or subtracting the 180 degrees. A simple mathematical mistake could have disastrous consequences.

---

1. Convert an azimuth into a back azimuth.
    a. Add 180 degrees if the azimuth is 180 degrees or less.
    b. Subtract 180 degrees if the azimuth is 180 degrees or more.
2. Convert a magnetic azimuth to a grid azimuth.
Note: To convert a magnetic azimuth to grid azimuth subtract G-M angle.

When converting azimuths, exercise extreme care when adding and subtracting the G-M angle. A simple mistake of one degree could be significant in the field.

    a. Locate the G-M angle, which is part of the declination diagram, at the bottom of your map. (Figure 1)

Note: The declination diagram shows both a graphic picture of the G-M angle and provides a written summary of the G-M angle. This written summary includes the date of the G-M angle and the actual angle in both degrees and mils.

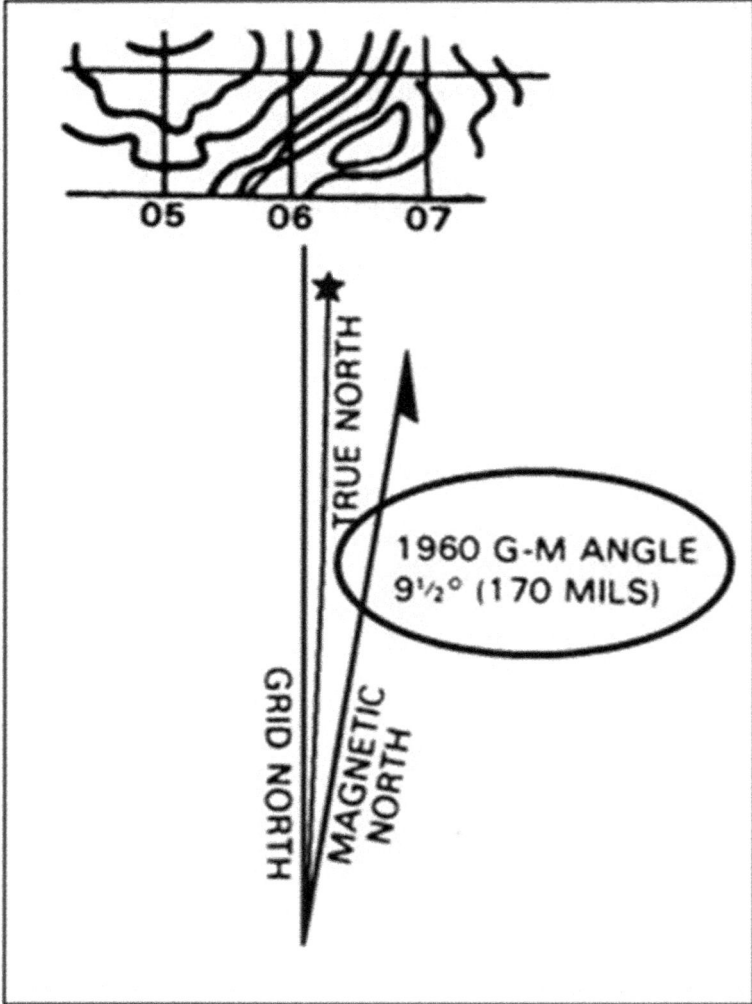

Figure 071-COM-0009-1. Declination diagram with G-M angle highlighted.

b. Determine whether the G-M angle is easterly or westerly. (figure 071-COM-0009-2)

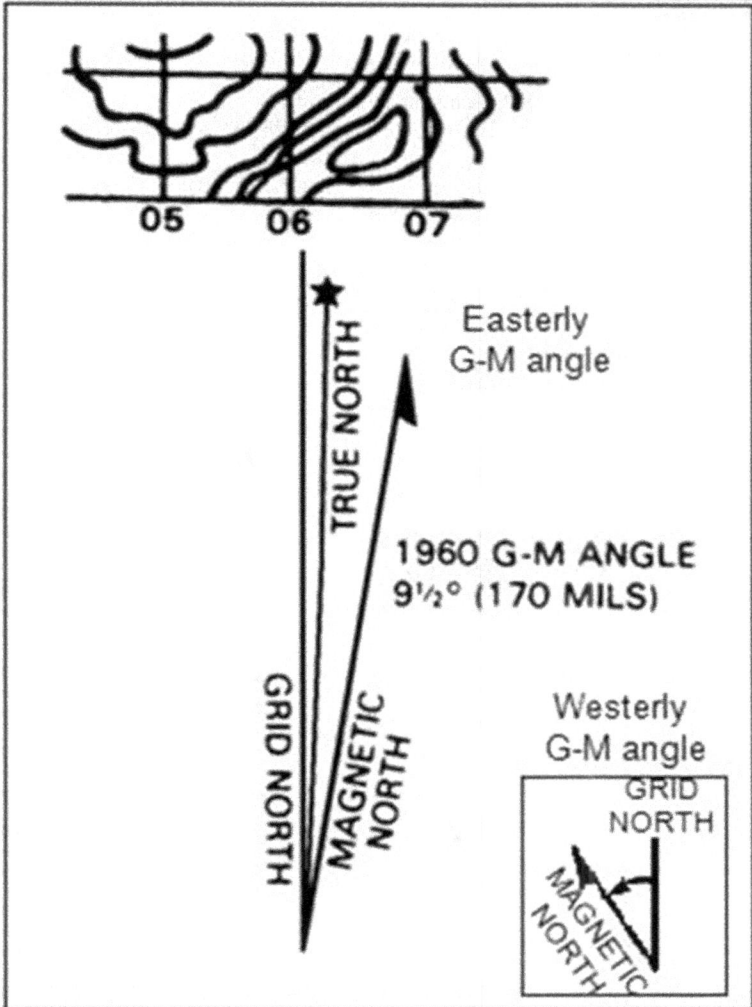

Figure 071-COM-0009-2. Identifying an easterly G-M angle and a westerly G-M angle.

c. Apply the mathematical formula to convert a magnetic azimuth to a grid azimuth.

(1) Add an easterly G-M angle to a magnetic azimuth to obtain a grid azimuth.

(2) Subtract a westerly G-M angle from a magnetic azimuth to obtain a

grid azimuth.

Note: As an example, given a magnetic azimuth of 190 degrees, using the above diagram you would add 9.5 degrees to 190 degrees to obtain a grid azimuth of 199.5 degrees.

3. Convert a grid azimuth to a magnetic azimuth.

      a. Locate the G-M angle, which is part of the declination diagram, at the bottom of your map. (Figure 071-COM-0009-3)

Note: The declination diagram shows both a graphic picture of the G-M angle and provides a written summary of the G-M angle. This written summary includes the date of the G-M angle and the actual angle in both degrees and mils.

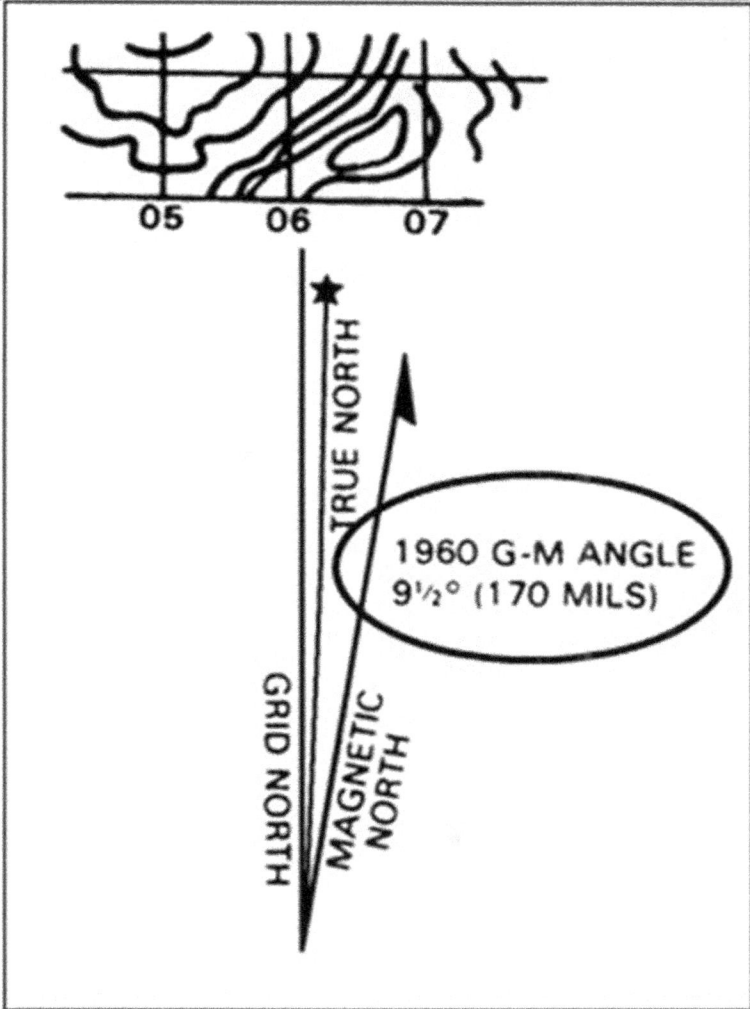

Figure 071-COM-0009-3. Declination diagram with G-M angle highlighted.

b. Determine whether the G-M angle is easterly or westerly. (Figure 071-COM-0009-4)

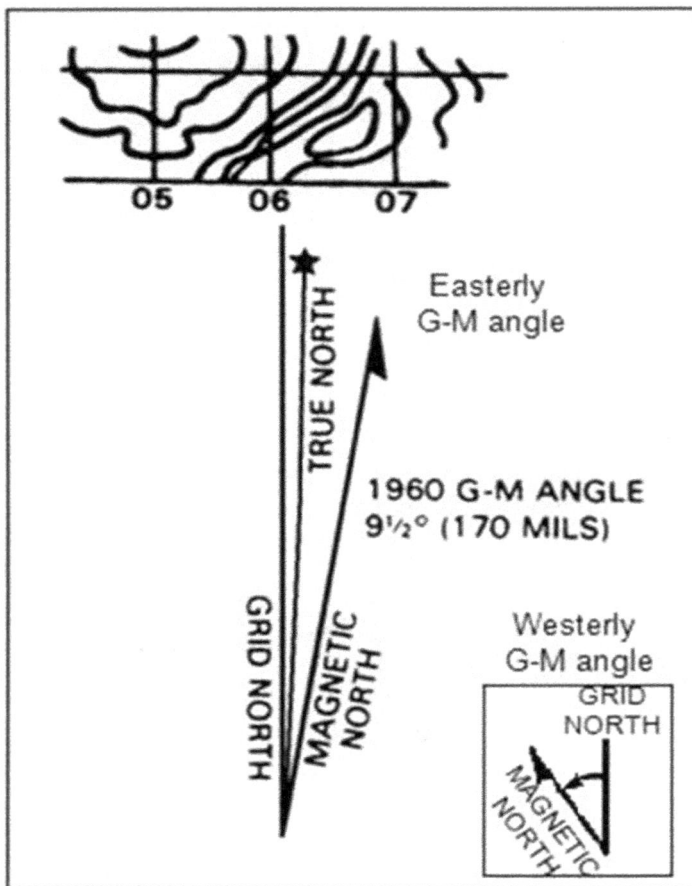

Figure 071-COM-0009-4. Identifying an easterly G-M angle and a westerly G-M angle.

   c. Apply the mathematical formula to convert a magnetic azimuth to a grid azimuth.

   (1) Subtract an easterly G-M angle from a grid azimuth to obtain a magnetic azimuth.

   (2) Add a westerly G-M angle to a grid azimuth to obtain a magnetic azimuth.

Note: As an example, given a grid azimuth of 199.5 degrees, using the above diagram you would subtract 9.5 degrees from 199.5 degrees to obtain a magnetic azimuth of 190 degrees.

(Asterisks indicates a leader performance step.)

Evaluation Guidance: Score the Soldier GO if all performance measures are passed. Score the Soldier NO-GO if any performance measure is failed. If the Soldier scores a NO-GO, show the Soldier what was done wrong and how to do it correctly.

Evaluation Preparation:

SETUP: Provide the Soldier with the equipment and/or materials described in the conditions statement.

BRIEF THE SOLDIER: Tell the Soldier what is expected by reviewing the task standards. Stress to the Soldier the importance of observing all cautions, warnings, and dangers to avoid injury to personnel and, if applicable, damage to equipment.

Performance Measures	GO	NO GO
1 Converted an azimuth into a back azimuth.	_____	_____
2 Converted a magnetic azimuth to a grid azimuth.	_____	_____
3 Converted a grid azimuth to magnetic azimuth.	_____	_____

Environment: Environmental protection is not just the law but the right thing to do. It is a continual process and starts with deliberate planning. Always be alert to ways to protect our environment during training and missions. In doing so, you will contribute to the sustainment of our training resources while protecting people and the environment from harmful effects. Refer to the current Environmental Considerations manual and the current GTA Environmental-related Risk Assessment card

Safety: In a training environment, leaders must perform a risk assessment in accordance with current Risk Management Doctrine. Leaders will complete the current Deliberate Risk Assessment Worksheet in accordance with the TRADOC Safety Officer during the planning and completion of each task and sub-task by assessing mission, enemy, terrain and weather, troops and support available-time available and civil considerations, (METT-TC). *Note:* During MOPP training, leaders must ensure personnel are monitored for potential heat injury. Local policies and procedures must be followed during times of increased heat category in order to avoid heat related injury. Consider the MOPP work/rest cycles and water replacement guidelines IAW current CBRN doctrine.

071-COM-0017
Compute Back Azimuths

Conditions: You are a member of a squad or team that is preparing for a mission and you have been directed to determine back azimuth(s) for given azimuth(s). The azimuths may be in either degrees or mils. Some iterations of this task should be performed in MOPP 4.

Standards: Determine the back azimuth for each given azimuth to the exact degree or mil.

Special Condition: None

Special Standards: None

Safety Risk: Low

MOPP 4: Sometimes

Task Statements

Cue: None

Note: None

WARNING

When converting azimuths into back azimuths, a simple mathematical mistake may cause disastrous consequences.

Remarks: None

Notes: None

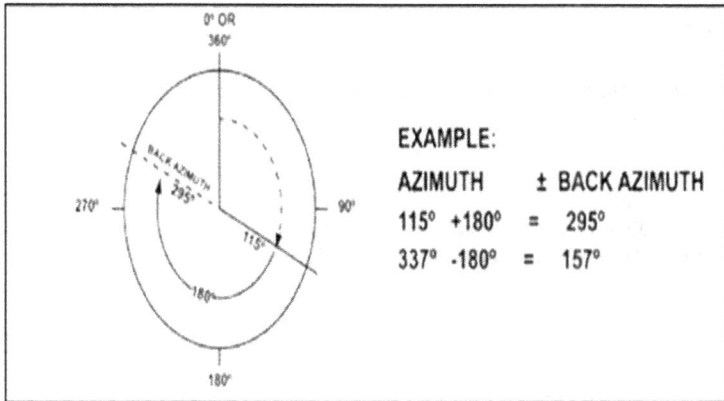

Figure 071-COM-0017-1. Calculation of a back azimuth.

Performance Steps

1. Determine back azimuth using degrees.

Note: Figure 071-COM-0017-1 provides an example on calculating a back azimuth in degrees.

    a. Add 180 degrees if the azimuth is 180 degrees or less.

    b. Subtract 180 degrees if the azimuth is 180 degrees or more.

2. Determine back azimuth using mils.

    a. If the azimuth is less than 3200 mils, add 3200 mils.

Note: As an example, given an 1150-mil azimuth, add 3200 to 1150 to obtain a back azimuth of 4350-mils. Mathematically, this is 1150 + 3200 = 4350.

    b. If the azimuth is more than 3200 mils, subtract 3200 mils.

Performance Measures	GO	NO GO
1 Determined back azimuth using degrees.	_____	_____
2 Determined back azimuth using mils.	_____	_____

References:
Required:
Related: TC 3-25.26

071-COM-1006

Navigate from One Point on the Ground to another Point while Dismounted

Conditions: You are a member of a squad or team in a field environment and have been directed to conduct movement to a designated point. You have a 1:50,000-scale topographic map of the area, a coordinate scale, a protractor, and a magnetic compass. Some iterations of this task should be performed in MOPP 4.

Standards: Navigate to the designated point using terrain association, dead reckoning, or a combination of both.

Special Condition: None

Special Standards: None

Safety Risk: Low

MOPP 4: Sometimes

Task Statements

Cue: None

Note: None

Performance Steps

1. Navigate using terrain association.
 a. Identify the start point and destination point on the map.
 b. Analyze the terrain between these two points for both movement and tactical purposes.
 c. Identify terrain features that can be recognized during movement, such as hilltops, roads, rivers, etc.
 d. Plan the best route, including checkpoints, if needed.
 e. Determine the map distances between identified checkpoints and the total distance to be traveled.
 f. Determine the actual ground distance by adding 20 percent to the map distance.
Note: Twenty percent is a general rule of thumb for cross country terrain - road movement and flat terrain do not require this 20 percent increase.
 g. Move to the designated end point (or intermediate point) using identified terrain features as aiming points or handrails.

Performance Steps

Note: Handrails are linear features like roads or highways, railroads, power transmission lines, ridgelines, or streams that run roughly parallel to your direction of travel.

    2. Navigate using dead reckoning.

Note: The use of steering marks is recommended when navigating by dead reckoning. A steering mark is a distant feature visible along one's route that is used as distant aiming point that one moves towards. Once reached another steering point is identified until a change of direction or the final destination is reached.

    a. Identify the start point and destination point on the map.

    b. Analyze the terrain between these two points for both movement and tactical purposes.

    c. Plan the best route, including checkpoints, if needed.

    d. Determine the grid azimuths between identified checkpoints (if any) and the final point.

    e. Convert the grid azimuth(s) taken from the map to a magnetic azimuth(s).

    f. Determine the map distances between identified checkpoints and the total distance to be traveled.

    g. Determine the direction of movement using the compass.

    h. Move in the identified direction of travel or towards the identified steering mark.

    i. Determine a new steering mark or confirm direction of travel as needed.

Note: The direction of movement, when not using a steering mark, must be periodically confirmed.

    3. Navigate using a combination of dead reckoning and terrain association.

    a. Follow the procedures outlined for both techniques.

    b. Use each technique to reinforce the accuracy of the other technique.

Evaluation Preparation:

Setup: Provide the Soldier with the equipment and or materials described in the conditions statement.

Brief Soldier: Tell the Soldier what is expected of him by reviewing the task standards. Stress to the Soldier the importance of observing all cautions, warnings, and dangers to avoid injury to personnel and, if applicable, damage to equipment.

Performance Measures	GO	NO GO
1 Navigated using terrain association.	_____	_____
2 Navigated using dead reckoning.	_____	_____

Performance Measures	GO	NO GO
3 Navigated using a combination of dead reckoning and terrain association.	_____	_____

Evaluation Guidance: Score the Soldier GO if all performance measures are passed. Score the Soldier NO-GO if any performance measure is failed. If the Soldier scores a NO-GO, show the Soldier what was done wrong and how to do it correctly.

References:
Required:
Related: TC 3-25.26

071-COM-1014

Locate an Unknown Point on a Map and on the Ground by Intersection

Conditions: You are a member of a squad or section and have a requirement to determine the location of the unknown point on the map. You have a 1:50,000-scale military map, a magnetic compass, a military protractor, pencil, paper, and an item that can be used as a straight edge. There are at least two well-defined points on the ground that you can locate on the map. Some iterations of this task should be performed in MOPP 4..

Standards: Determine the grid coordinates of the unknown point to within 100 meters; include the two-letter 100,000 meter square identifier, using either the map-and-compass method or the straight edge method.

Special Standards: None

Safety Risk: Low

MOPP 4: Sometim

Note: Intersection is the location of an unknown point by occupying at least two (preferably three) known positions on the ground (either successively by one Soldier or simultaneously by two or more Soldiers), then plotting on the map the grid azimuth of each of these known points to the unknown point, and identifying the point on the map where the lines intersect. It is used to locate distant or inaccessible points or objects such as enemy targets and danger areas. There are two methods of intersection: the map-and-compass method and the straight edge method.

Performance Steps

1. Identify an unknown point on a map by intersection using the map-and-compass method. (Figure 071-COM-1014-1)

    a. Orient the map on a flat surface using a compass.

    b. Plot grid azimuths from known points to the unknown point on the map.

        (1) Mark your position (the observers) on the map.

        (2) Determine the magnetic azimuth from your position to the unknown point.

        (3) Convert the magnetic azimuth to a grid azimuth.

        (4) Place the index point of a protractor on your plotted position.

        (5) Align the protractor's 0 to 180-degree line to the top of the map's North-South grid line.

        (6) Ensure the 0-degree mark is pointing to the north (or top of map).

        (7) Place a tick mark on the map beside the number on the protractor that corresponds to the computed grid azimuth.

        (8) Draw a straight line from your plotted position to the tick mark and beyond.

        (9) Repeat steps 1b(1) through 1b(8) for each observer position.

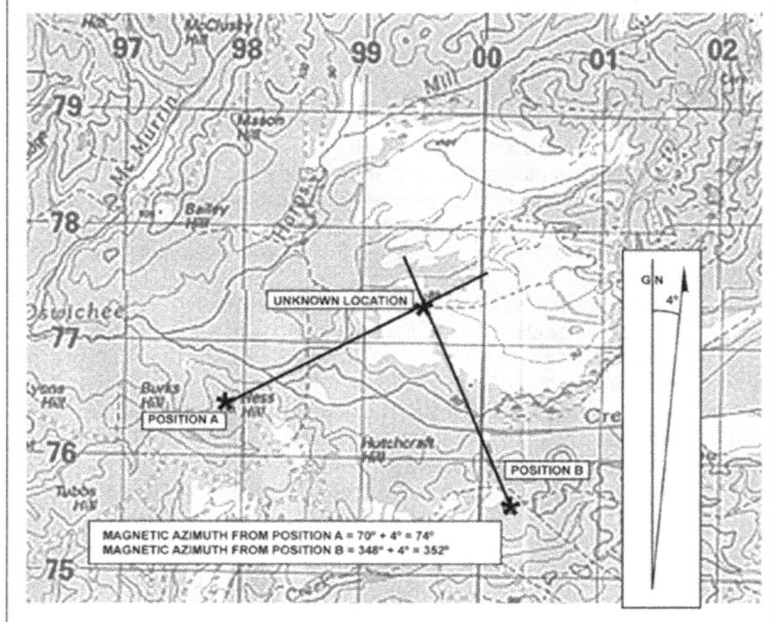

Figure 071-COM-1014-1.Intersection Using the Map-and-Compass Method

(10) Identify the point where the lines intersect as the location of the unknown point.

(11) Determine the grid coordinates to this location to the desired accuracy.

2. Identify an unknown point on a map by intersection using the straight edge method. (Figure 071-COM-1014-2)

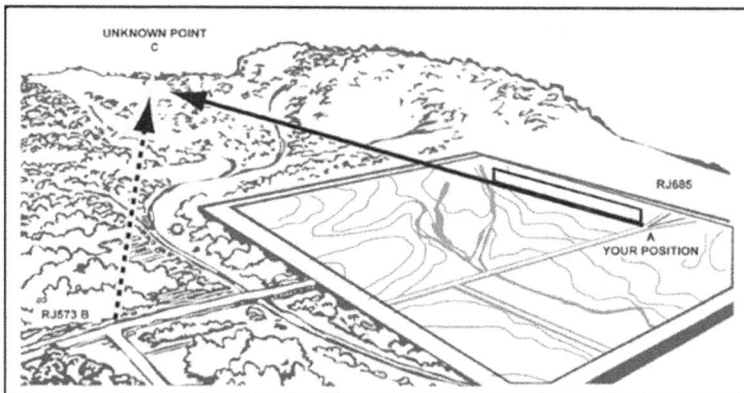

Figure 071-COM-1014-2. Intersection using the Straight Edge Method

    a.   Orient your map on a flat surface using terrain association.

    b.   Mark your position (the observers) on the map.

    c.   Draw an intersection line for each of these plotted points.

        (1) Lay a straight edge on one of the two known observer points on the map.

        (2) Rotate the straightedge on the map until the straightedge lines up with both the known observer position on the map (Point A and Point B in Figure 2) and the unknown position in the distance (Point C in Figure 2).

        (3) Draw a line along the straight edge from the known observer position toward the unknown position on the ground.

        (4) Repeat steps 2c(1) through 2c(3) for each plotted point.

    d.   Identify the point where the lines intersect as the unknown location.

    e.   Determine the grid coordinates to this location to the desired accuracy. (Asterisks indicates a leader performance step.)

Evaluation Preparation:

Setup: Provide the Soldier with the equipment and/or materials described in the conditions statement.

Brief Soldier: Tell the Soldier what is expected by reviewing the task standards. Stress to the Soldier the importance of observing all cautions, warnings, and dangers to avoid injury to personnel and, if applicable, damage to equipment.

Performance Measures	GO	NO GO
1 Identified an unknown point on a map by intersection using the map-and-compass method.	_____	_____
2 Identified an unknown point on a map by intersection using the straight edge method.	_____	_____

Evaluation Guidance: Score the Soldier GO if all performance measures are passed. Score the Soldier NO-GO if any performance measure is failed. If the Soldier scores a NO-GO, show the Soldier what was done wrong and how to do it correctly.

Environment: Environmental protection is not just the law but the right thing to do. It is a continual process and starts with deliberate planning. Always be alert to ways to protect our environment during training and missions. In doing so, you will contribute to the sustainment of our training resources while protecting people and the environment from harmful effects. Refer to the current Environmental Considerations manual and the current GTA Environmental-related Risk Assessment card.

Safety: In a training environment, leaders must perform a risk assessment in accordance with current Risk Management Doctrine. Leaders will complete the current Deliberate Risk Assessment Worksheet in accordance with the TRADOC Safety Officer during the planning and completion of each task and sub-task by assessing mission, enemy, terrain and weather, troops and support available-time available and civil considerations, (METT-TC). *Note:* During MOPP training, leaders must ensure personnel are monitored for potential heat injury. Local policies and procedures must be followed during times of increased heat category in order to avoid heat related injury. Consider the MOPP work/rest cycles and water replacement guidelines IAW current CBRN doctrine.

References:
Required:
Related: TC 3-25.26

071-COM-1015

Locate an Unknown Point on a Map and on the Ground by Resection

Conditions: You are a member of a squad or section and have a requirement to determine the teams current location. You have a 1:50,000-scale military map, a magnetic compass, a military protractor, pencil, paper, and an item that can be used as a straight edge. There are at least two well-defined points on the ground that you can locate on the map. Some iterations of this task should be performed in MOPP 4..

Standards: Determine the grid coordinates of the unknown point to within 100 meters; include the two-letter 100,000 meter square identifier, using either the map-and-compass method or the straight edge method.

Notes: Resection is the method of locating one's position on a map by determining the grid azimuth to at least two well-defined locations that can be pinpointed on the map. For greater accuracy, the desired method of resection is to use three or more well-defined locations.

Special Standards: None

Safety Risk: Low

MOPP 4: Sometimes

Task Statements

Cue: None

Remarks: None

Performance Steps

    1. Identify your location on a map by resection using the map and compass method (Figure 071-COM-1015-1).

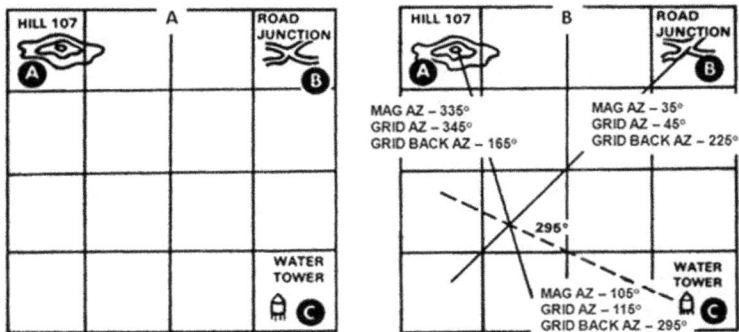

Figure 071-COM-1015-1. Resection using the map and compass method.

    a. Orient the map on a flat surface using a compass.
    b. Identify at least two well-defined points on the ground.

c. Mark these well-defined points on the map. (Figure 1, Example A).

d. Plot the back azimuths of these points on the map. (Figure 1, Example B)

(1) Determine the magnetic azimuth from your location to one of the defined points.

(2) Convert the magnetic azimuth to a grid azimuth.

(3) Convert this grid azimuth to a back grid azimuth.

(4) Place the index point of a protractor on the well-defined point.

(5) Align the protractor's 0- to 180-degree line to the top of the map's North-South grid line.

(6) Ensure the 0-degree mark is pointing to the north (or top of map).

(7) Place a tick mark on the map beside the number on the protractor that corresponds to the computed back grid azimuth.

(8) Draw a straight line from the well-defined point to the tick and beyond.

(9) Repeat steps 1d (1) through 1d (8) for each well-defined point.

e. Identify the point where the lines intersect as your location.

f. Determine the grid coordinates to this location to the desired accuracy.

2. Identify your location on a map by resection using the straightedge method (Figure 071-COM-1015-2).

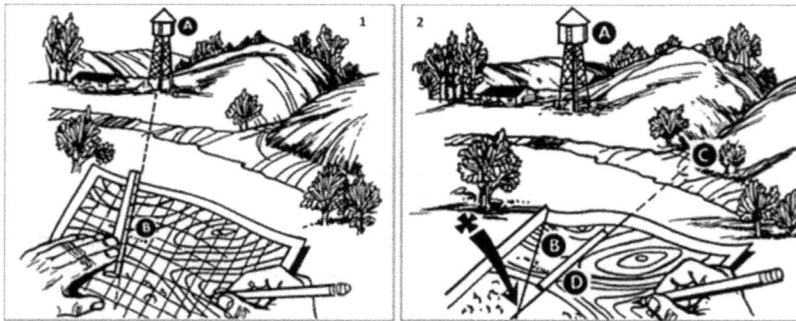

Figure 071-COM-1015-2. Resection using the straight edge method.

a. Orient your map on a flat surface using terrain association.

b. Locate at least two known distant locations or prominent features on the ground.

c. Plot these distant locations or prominent features on the map.

d. Draw a resection line for each of these plotted points.

(1) Lay a straightedge on one of the two known points on the map.

(2) Rotate the straightedge on the map until straight edge lines up with both the known position on the map (Figure 2, Point B and Point D) and the known position in the distance (Figure 2, Point A and Point C).

(3) Draw a line along the straightedge away from the known position on the ground toward your position.

(4) Repeat steps 2d (1) through 2d(3) for each plotted point.

e. Identify the point where the lines intersect as your location.

f. Determine the grid coordinates to this location to the desired accuracy.

(Asterisks indicates a leader performance step.)

Evaluation Preparation:

Setup: Provide the Soldier with the equipment and/or materials described in the conditions statement.

Brief Soldier: Tell the Soldier what is expected by reviewing the task standards. Stress to the Soldier the importance of observing all cautions, warnings, and dangers to avoid injury to personnel and, if applicable, damage to equipment.

Performance Measures	GO	NO GO
1 Identified your location on a map by resection using the map and compass method.	_____	_____
2 Identified your location on a map by resection using the straightedge method.	_____	_____

Evaluation Guidance: Score the Soldier GO if all performance measures are passed. Score the Soldier NO-GO if any performance measure is failed. If the Soldier scores a NO-GO, show the Soldier what was done wrong and how to do it correctly.

References:
Required:
Related: TC 3-25.26

071-COM-0501

Move as a Member of a Team

Conditions: You are a member of a dismounted team that is conducting tactical movement. You are not the team leader. You have your individual weapon and individual combat equipment. Some iterations of this task should be performed in MOPP 4.

Standards: Assume your position in the team's current formation, maintain proper distance between you and other team members, follow the team leader's example, and maintain security of your sector.

Special Condition: None

Special Standards: None

Safety Risk: Medium

MOPP 4: Sometimes

Task Statements

Cue: None

*Note:*The standard team is composed of four personnel - team leader (TL), automatic rifleman (AR), grenadier (G), and rifleman (R). The team leader designates positions based on the mission variables.

Performance Steps

1. Assume your position in the team's current formation.
Note: Specific positions vary based on the type of movement formation selected by the team leader.

   a. Assume your position within the team wedge formation (Figure 071-COM-0501-1).

Note: This is the basic team formation. It is easy to control, is flexible, allows immediate fires in all directions, and offers all-round local security.

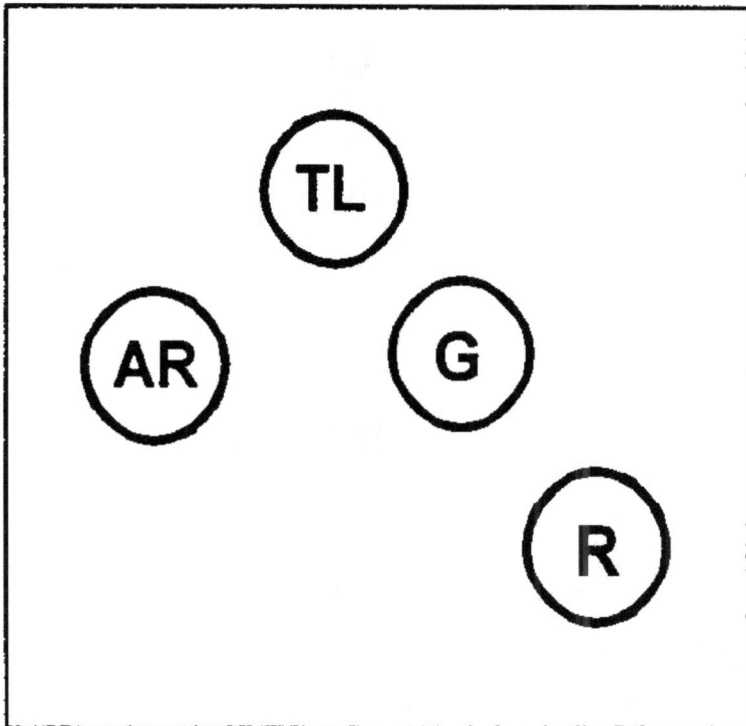

Figure 071-COM-0501-1. Wedge Formations

b. Assume your position within the team file formation (Figure 071-COM-0501-2).

Note: The file is used when employing the wedge is impractical. This formation is most often used in severely restrictive terrain, like inside a building; dense vegetation; limited visibility; and so forth. The distance between Soldiers changes due to constraints of the situation, particularly when in urban operations.

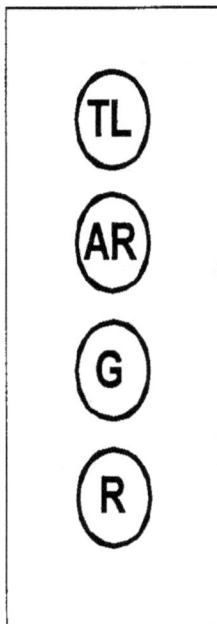

Figure 071-COM-0501-2. File Formation.

2. Maintain proper distance between you and other team members.
Note: The normal distance between Soldiers is 10 meters. When enemy contact is possible, the distance between teams should be about 50 meters. In open terrain such as desert, the interval may increase. The distance between individuals is determined by how much control the team leader can still exercise over his team members.

3. Maintain visual contact with your team leader.
Note: It is essential for all team members to maintain visual contact with the team leader.

4. Follow the team leader's example.
Note: When the team leader moves left, you move to the left. When the team leader gets down, you get down.

5. Adjust your position within the team as designated by the team leader.

6. Maintain security of your sector (i.e. to the flanks, front or rear of the team).

Evaluation Preparation:

Setup: Provide the Soldier with the equipment and or materials described in the conditions statement.

Brief Soldier: Tell the Soldier what is expected of him by reviewing the task standards. Stress to the Soldier the importance of observing all cautions, warnings, and dangers to avoid injury to personnel and, if applicable, damage to equipment.

Performance Measures	GO	NO GO
1 Assumed position in the team's current formation	_____	_____
2 Maintained proper distance from other team members.	_____	_____
3 Maintained visual contact with the team leader.	_____	_____
4 Followed the team leader's example.	_____	_____
5 Changed position within the team as designated by the team leader.	_____	_____
6 Maintained security of assigned sector.	_____	_____

Evaluation Guidance: Score the Soldier GO if all performance measures are passed. Score the Soldier NO-GO if any performance measure is failed. If the Soldier scores a NO-GO, show the Soldier what was done wrong and how to do it correctly.

References:
Required:
Related: TC 3-21.75

071-COM-0502

Move Under Direct Fire

Conditions: You are a member of a team conducting movement to contact and are under fire from an enemy position that is 250 to 300 meters away from your position. You have an individual weapon, individual combat equipment, and a current firing position that provides cover from the enemy's direct fire. Some interation of this task should be performed in MOPP 4.

Standards: Move within 100 meters of the enemy position using the appropriate movement techniques based on the situation and terrain.

Special Condition: None

Safety Risk: Medium

Special Equipment:

MOPP 4: Sometimes

Task Statements

Cue: None

*Note:*While this task may be performed by an individual Soldier, it is best performed as a member of a team or as part of a two-man buddy team.

Performance Steps

1. Select an individual movement route that adheres to the instructions provided by your team leader.
Note: When part of a team your movement route and general firing positions may be determined by your team leader. When moving as part of a team you must be prepared to follow your team leader's example.
    a. Search the terrain to your front for good firing positions.
Note: Large trees, rocks, stumps, fallen timber, rubble, vehicle hulls, man-made structures, and folds or creases on the ground may provide both cover and concealment and can be used as fighting positions.
    b. Select the best route to the positions.
Note: A gully, ravine, ditch, or wall at a slight angle to your direction of travel may provide cover and concealment when using the low or high crawl movement techniques. Hedge rows or a line of thick vegetation may provide concealment only when using the low or high crawl technique.

(1) Pick a route that minimize your exposure to enemy fire.

(2) Ensure route does not cross in front of other team members.

2. Communicate your movement intent to your buddy and team leader, as appropriate, using hand and arm signals.

3. Suppress the enemy as required.

Note: Do not expose yourself to fire unless the enemy is suppressed. Suppression of the enemy may be accomplished by another element, a buddy, or by yourself. With the enemy suppressed you can select an individual movement route or initiate movement.

4. Conduct movement using the appropriate technique(s) to reach each position.

a. Move using the high crawl technique (figure 071-COM-1502-1).

Note: The high crawl lets you move faster than the low crawl and still gives you a low silhouette. Use this crawl when there is good cover and concealment but enemy fire prevents you from getting up.

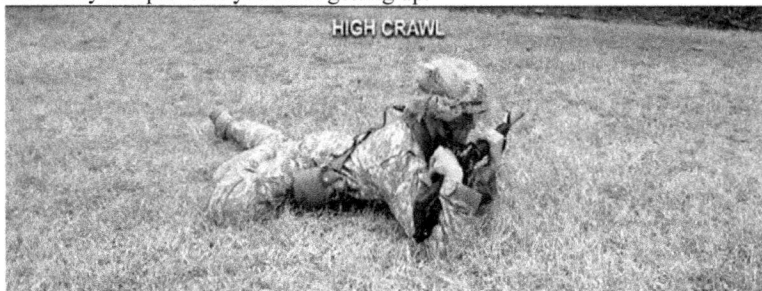

Figure 071-COM-1502-1. High Crawl.

(1) Keep your body off of the ground.

(2) Rest your weight on your forearms and lower legs.

(3) Cradle your weapon in your arms.

(4) Keep the muzzle of the weapon off the ground.

(5) Keep your knees well behind your buttocks so it stays low.

(6) Move forward by alternately advancing your right elbow and left knee, and left elbow and right knee.

b. Move using the low crawl technique (figure 071-COM-1502-2).

Note: The low crawl gives you the lowest silhouette. It is used to cross places where the cover and/or concealment are very low and enemy fire or observation prevents you from getting up.

Figure 071-COM-1502-2. Low Crawl.

(1) Keep your body as flat as possible to the ground.

(2) Grasp the sling of the weapon at the upper sling swivel with your right hand.

(3) Let the hand guard rest on your forearm.

(4) Keep the muzzle of the weapon off the ground.

(5) Move forward.

(a) Push both arms forward while pulling your right leg forward.

(b) Pull on the ground with both arms while pushing with your right leg.

(c) Repeat steps (a) and (b) until you reach your next position.

c. Moved using the rush technique (figure 071-COM-1502-3).

Note: The rush is the fastest way to move from one position to another. Use when you must cross an open area and time is critical.

Figure 071-COM-1502-3. Rush.

(1) Raise your head.
(2) Select your next position.
(3) Lower your head.
(4) Draw your arms into your body.
(5) Pull your right leg forward.
(6) Raise your body.
(7) Get up quickly.
(8) Run for 3-5 seconds to your next position.
(9) Plant both feet just before hitting the ground.
(10) Fall forward.
    (a) Drop to your knees.
    (b) Slide your right hand down to the heel of the butt of your weapon.
    (c) Break your fall with the butt of your weapon.
   d. Continue using movement techniques until you reach your final firing position.

Performance Steps

5. Occupy your identified firing position within 100 meters of the enemy position.
   a. Assume a firing position.
   b. Engage enemy with your individual weapon.

Evaluation Preparation:

Setup: Provide the Soldier with the equipment and/or materials described in the conditions statement.

Brief Soldier: Tell the Soldier what is expected of him by reviewing the task standards. Stress to the Soldier the importance of observing all cautions, warnings, and dangers to avoid injury to personnel and, if applicable, damage to equipment.

Performance Measures	GO	NO GO
1 Selected an individual movement route that adhered to the instructions provided by your team leader.	_____	_____
2 Communicated movement intent to buddy and team leader, as appropriate, using hand and arm signals.	_____	_____
3 Suppressed the enemy as required.	_____	_____
4 Conducted movement using the appropriate technique(s) to reach each position.	_____	_____
5 Occupied your identified firing position within 100 meters of the enemy position.	_____	_____

Evaluation Guidance: Score the Soldier GO if all performance measures are passed. Score the Soldier NO-GO if any performance measure is failed. If the Soldier scores a NO-GO, show the Soldier what was done wrong and how to do it correctly.

References:
Required:
Related: TC 3-21.75

071-COM-0510

React to Indirect Fire while Dismounted

Conditions: You are a member of a squad or team conducting a dismounted patrol and you hear indirect fire exploding or passing Over-head. You have your individual weapon and equipment. Some iterations of this task should be performed in MOPP 4.

Standards: React to indirect fire while moving as a member of a squad or team.

Special Condition: None

Safety Risk: Medium

MOPP 4: Sometimes

Task Statements

Cue: None

Performance Steps

1. Shout "Incoming!" in a loud, recognizable voice.
2. Drop to the ground.
3. Follow commands and actions of your leader.

Note: Normally, if moving, the leader will tell you to run out of the impact area in a certain direction or will tell you to follow him. If you cannot see or hear your leader you should follow other team members.

4. Seek the nearest appropriate cover.
5. Avoid the impact area if not already in it.
6. Run in a direction away from the incoming fire.
7. Assess your situation.
8. Report your situation to your leader.
9. Continue the mission.

Evaluation Preparation:

Setup: Provide the Soldier with the equipment and/or materials described in the conditions statement.

Brief Soldier: Explain what is expected from the Soldier by reviewing the task standards. Stress the importance of observing all cautions, warnings, and dangers to avoid injury to personnel and, if applicable, damage to equipment.

Performance Measures	GO	NO GO
1 Shouted "Incoming!" in a loud, easily recognizable voice.	_____	_____
2 Drop to the ground.	_____	_____
3 Followed the commands and actions of your leader.	_____	_____
4 Seeked the nearest appropriate cover.	_____	_____
5 Avoided the impact area if not already in it.	_____	_____
6 Ran in direction away from the incoming fire.	_____	_____
7 Assessed your situation.	_____	_____
8 Reported your situation to your leader.	_____	_____
9 Continued the mission.	_____	_____

Evaluation Guidance:. Refer to chapter 1, paragraph 1-9e, (1) and (2).

References: TC 3-21.75

071-COM-0513

Select Hasty Fighting Positions

Condition: You are a member of a dismounted squad or team occupying an area and have been directed to establish a temporary fighting position to cover a given sector of fire. You have an individual or crew-served weapon and your individual combat equipment.

Standard: Select and prepare a hasty fighting position that protects you from enemy observation and fire, and allows effective fires to be placed within sector of fire.

Special Condition: None

Safety Risk: Low

MOPP 4:

Task Statements

Performance Steps

1. Identify a position that will provide the best cover and concealment.

Note: Cover, made of natural or man-made materials, gives protection from bullets, fragments of exploding rounds, flame, nuclear effects, biological and chemical agents, and enemy observation. Concealment is anything that hides personal, equipment and/or vehicles from enemy observation. Concealment does not protect you from enemy fire.
    a. Use natural, undisturbed cover and concealment, if available.
    b. Ensure man-made cover and concealment blends with surroundings.
2. Ensure the position allows effective weapon emplacement.
    a. Ensure proper sector of fires for appropriate weapon system.
    b. Ensure proper field of fires.
3. Prepare the fighting position.
    a. Avoid disclosing your position by careless or excessive clearing.
    b. Leave a thin, natural screen of vegetation to hide your position.
    c. Cut off lower branches of large, scattered trees, in sparsely wooded areas.
    d. Clear underbrush only where it blocks your view.
    e. Remove cut brush, limbs, and weeds so the enemy will not spot them.
    f. Cover cuts on trees and bushes forward of your position with mud, dirt, or snow.
    g. Leave no trails as clues for the enemy.
4. Maintain camouflage.
Note: Camouflage is anything you use to keep yourself, your equipment, and your position from being identified.
    a. Prevent attention by controlling movement and activities.
    b. Avoid putting anything where the enemy expects to find it.
    c. Break up outlines and shadows.
    d. Conceal shining objects.
    e. Break up familiar shapes to make them blend in with their surroundings.

f. Camouflage yourself and your equipment to blend with the surroundings.

g. Ensure proper dispersion.

h. Study the terrain and vegetation of the area in which you are operating.

i. Use camouflage material that best blends with the area.

Evaluation Preparation:

Setup: Provide the Soldier with the equipment and/or materials described in the conditions statement.

Brief Soldier: Tell the Soldier what is expected by reviewing the task standards. Stress to the Soldier the importance of observing all cautions, warnings, and dangers to avoid injury to personnel and, if applicable, damage to equipment

Performance Measures	**GO**	**NO GO**
1 Identified a position that provided the best cover and concealment.	_____	_____
2 Ensured the position allowed effective weapon emplacement.	_____	_____
3 Prepared the fighting position.	_____	_____
4 Maintained camouflage.	_____	_____

Evaluation Guidance: Score the Soldier GO if all performance measures are passed. Score the Soldier NO-GO if any performance measure is failed. If the Soldier scores a NO-GO, show the Soldier what was done wrong and how to do it correctly.

Environment: Environmental protection is not just the law but the right thing to do. It is a continual process and starts with deliberate planning. Always be alert to ways to protect our environment during training and missions. In doing so, you will contribute to the sustainment of our training resources while protecting people and the environment from harmful effects. Refer to ATP 3-34.5 Environmental Considerations and GTA 05-08-002 ENVIRONMENTAL-RELATED RISK ASSESSMENT. Environmental protection is not just the law but the right thing to do. It is a continual process and starts with deliberate planning. Units will assess environmental risk using the assessment matrixes in ATP 3-34.5, Appendix D. Always be alert to ways to protect our environment during training and missions. In doing so, you will contribute to the sustainment of our training resources while protecting people and the environment from harmful effects.

Safety: In a training environment, leaders must perform a risk assessment in accordance with ATP 5-19, Risk Management. Leaders will complete a DD Form 2977 DELIBERATE RISK ASSESSMENT WORKSHEET during the planning and completion of each task and sub-task by assessing mission, enemy, terrain and weather, troops and support available-time available and civil considerations, (METT-TC). *Note:* During MOPP training, leaders must ensure personnel are monitored for potential heat injury. Local policies and procedures must be followed during times of increased heat category in order to avoid heat related injury. Consider the MOPP work/rest cycles and water replacement guidelines IAW TM 3-11.32 Multi-Service Reference For Chemical, Biological, Radiological, and Nuclear Warning And Reporting And Hazard Prediction Procedures.

References:
Required:
Related: TC 3-21.75

Subject Area 4: Communicate

113-COM-2070

Operate Single Channel Ground and Airborne Radio System SINCGARS

Conditions: Given a requirement to contact a distant radio station, an operational SINCGARS, loaded communication security (COMSEC) fill device, antenna, distant station, Signal Operations Instructions (SOI), and required references in an operational environment. Some iterations of this task should be performed in MOPP 4.

Standards: Prepare SINCGARS in Single Channel (SC) and Frequency Hopping (FH) modes; establish communications with a distant station by successfully conducting a secure communications check according to required references.

Special Condition: None

Safety Risk: Low

MOPP 4: Sometimes

Task Statements

DANGER

Unsafe procedures may cause loss of life or damage to equipment. Personnel in units must be familiar with the content of all pertinent publications. Technical bulletins and technical manuals provide information on the safe handling, use, storage, and maintenance requirements of tools, equipment, and hazardous materials. Shortcuts or deviations can result in accidents.

Cue: You may begin.

WARNING

Service members may have to depend upon their first aid knowledge and skills to save themselves or other service members. Service members may be exposed to hazardous material and or electrical shock. Service members need to know what to do, what not to do, and when to seek medical assistance. Follow all safety procedures IAW TM 11-5820-890-13&P-1. For more information, refer to TC 4-02.1, First Aid for Soldiers.

CAUTION

Identifying hazards and controlling risks across the full spectrum of Army functions, operations, and activities is the responsibility of all Soldiers. This equipment contains parts that are susceptible to damage by Electrical Static Discharge (ESD). Follow all proper handling and installation procedures IAW TM 11-5820-890-13&P-1. Leaders alert all users to the possibility of personal injury or damage to equipment that may result from long- term failure to follow correct procedures and to avoid all unsafe acts.

Remarks: None
Notes: Stop time or evaluation between modes of operation.

Performance Steps
    1. Operate SINCGARS in SC mode.

    a. Perform start-up procedures.

    b. Enter the following parameters.

    (1) Turn FCTN knob to Load (LD).

    (2) Put radio into SC MODE.

Note: For R/T-1523D and previous models turn the MODE knob to SC. For R/T-1523E, and later models press the MENU button repeatedly to navigate to the MODE options; then press the CHG button to change the radio to SC.

    (3) Select Channel (CHAN) 1-6, CUE, or MAN.

Note: For R/T-1523D and pervious models use the CHAN knob to change channels. For R/T-1523E and later models use the MENU button to navigate to CHANNEL and press button 1-6, CUE (0), or MAN (7) for correlating channel.

    (4) Press Frequency (FREQ), then Clear (CLR).

    (5) Enter frequency.

    (6) Press Store (STO).

    (7) Repeat the above steps b3-b6 for each channel requiring a frequency

    c. Load Communication Security (COMSEC).

    (1) Change COMSEC MODE to Cypher Text (CT).

Note: For R/T-1523D, and previous models turn the COMSEC knob to CT. For R/T-1523E, and later models press the MENU button repeatedly to navigate to the COMSEC option menu; then press the CHG button to change the radio to CT.

    (2) Connect COMSEC fill device to AUD/FILL port.

    (3) Initiate COMSEC transfer by pressing LOAD (0) then STO and then the corresponding channel key.

Note: Repeat for each channel requiring COMSEC.

    d. Turn FCTN knob to Squelch (SQ)/ON.

    e. Conduct a radio check.

    (1) Enter the net using proper radio procedures IAW ACP-125 SUPP1.

    (2) Establish communication with distant end.

  2. Operate SINCGARS in Frequency Hop (FH) mode.

    a. Perform start-up procedures.

    b. Execute the FH load procedures.

    (1) Turn FCTN knob to LD.

    (2) Put radio into FH MODE.

Note: For R/T-1523D, and previous models turn the MODE knob to FH. For R/T-1523E, and later models press the MENU button repeatedly to navigate to the MODE options; then press the CHG button to change the radio to FH.

    (3) Put radio into CT mode.

Note: For R/T-1523D, and previous models turn the COMSEC knob to CT. For R/T-1523E, and later models press the MENU button repeatedly to navigate to the COMSEC options; then press the CHG button to change the radio to CT.

    c. Load COMSEC.

    (1) Connect COMSEC fill device to AUD/FILL port.

    (2) Initiate COMSEC transfer by press LOAD (0) then STO then the correlating channel key.

Note:

(3) Load the HOPSET.

Note: HOPSET loading procedures vary greatly between fill devices; reference the appropriate manual for the available fill device.

d. Set the radio date.

Note: Press TIME, then MENU/CLR. Enter current Julian date then press STO. Reference DA PAM 738–750 for Julian date.

e. Set radio time.

Note: Press TIME, then press MENU/CLR. Enter current Zulu/Universal Time Co-ordinated (UTC; press STO. The time starts counting as soon as the STO button is pressed. Accuracy in timing is essential to SINCGARS operations, deviations of +4 seconds could result in inability to communicate.

f. Turn FCTN knob to SQ/ON.

g. Conduct a radio check (repeat step 1e).

Evaluation Preparation:

Performance Measures	GO	NO GO
1 Prepared SINCGARS for secure SC operations.	_____	_____
2 Conduct secure radio check in SC mode with distant station.	_____	_____
3 Conduct secure radio check in FH mode with distant station.	_____	_____
4 Performing stopping procedures.	_____	_____

Evaluation Guidance: Score the soldier GO if all steps are passed. Score the soldier NO-GO if any step is failed. If the soldier fails any step, show what was done wrong and how to do it correctly. Have the soldier practice until the task can be performed correctly.

References
Required: ACP 125, TM 11-5820-890-13&P-1, TM 11-5820-890-13&P-2, TM 11-5820-890-13&P-3, TM 11-5820-890-13&P-4, TM 11-5820-890-13&P-5, TM 11-5820-890-13&P-6, TM 11-5820-890-13&P-7, TM 11-5820-890-13&P-8, TM 11-5820-890-13&P-9, TM 11-5820-890-13&P-10, UNIT SOI Unit/Unit's Signal Operation Instructions (SOI)

References

113-COM-1022

Perform Voice Communications

Conditions: Given: 1. One operational radio set for each net member, warmed up and set to the net frequency. 2. A call sign information card (5 x 8) consisting of: Net member duty position (S-1, S-2), net call sign (letter-number-letter), suffix list (Net Control Station [NCS] - 46, S-1 - 39, S-2 - 13), and a message to be transmitted. 3. Situation: The net is considered to be secure and authentication is not required. *Note:* This task may have as many net members as there is equipment available. Each net member must have a different suffix and message to transmit.

Standards: Perform voice communications by entering and leaving the net in alphanumeric sequence and use of correct prowords, phonetic alphabet, and numerals that result in successful communication of the message.
Special Condition: None

Special Standards: None

Safety Risk: Low

Task Statements

Cue: None

Note: None

Performance Steps

1. Enter the net.
    a. Determine the abbreviated call sign and answering sequence for your duty position.
    b. Respond to the NCS issuing a net call.
    c. Answer in alphanumeric sequence.
Note: At this time, the NCS acknowledges and the net is open.
2. Send a message.
    a. Listen to make sure the net is clear. Do not interrupt any ongoing communications.
    b. Call the NCS and tell the operator the priority of the message you have for his or her station.

c. Receive a response from the NCS that he or she is ready to receive.

d. Send your message using the correct prowords and pronunciation of letters and numbers.

e. Get a receipt for the message.

3. Leave the net in alphanumeric sequence.

Note: The NCS acknowledges and the net is closed. *Note:* The following call signs are used in this task as an example: Net call sign - E3E, NCS - E46, S-1 - E39, S-2 - E13.

a. Answer in alphanumeric sequence.

b. You receive a call from the NCS who issues a close down order.

Evaluation Preparation:

Setup: Position operational radio sets in different rooms or tents or at least 70 feet apart outside. Obtain call signs, suffixes, and a radio frequency through the normal command chain. Select a message 15-25 words in length, containing some number groups such as map coordinates and times. Print the call signs for the sender and the receiver, along with the message to be sent, on 5 x 8 cards. Perform a communications check to ensure operation of the radios. Have an assistant who is proficient in radio operation man the NCS. Provide the assistant with the call signs. If the soldier has not demonstrated sufficient progress to complete the task within 5 minutes, give him or her a NO-GO. This time limit is an administrative requirement, not a doctrinal one; so if the soldier has almost completed the task correctly, you may decide to allow him or her to finish.

Brief Soldier: Give the soldier to be tested the card containing the message and call signs. Tell him or her the radio is ready for operation, the net is considered to be secure and authentication is not required, and to send the message to the NCS and get a receipt. Tell the soldier, if sufficient progress in completing the task within 5 minutes has not been demonstrated, he or she will receive a NO-GO for the task.

Performance Measures	GO	NO GO
1 Entered the net in alphanumeric sequence.	_____	_____
2 Sent a message of 15 to 25 words using the correct prowords and phonetic alphabet and numerals.	_____	_____
3 Left the net in alphanumeric sequence.	_____	_____

Evaluation Guidance: Score the soldier GO if all performance measures are passed. Score the soldier NO-GO if any performance measure is failed. If the soldier scores NO-GO, show the soldier what was done wrong and how to do it correctly.

References:
Required:
Related: ACP 125 (F); ACP 131 (E);
TB 9-2320-280-35-2

081-COM-0101

Request Medical Evacuation

Conditions: You have a casualty requiring medical evacuation (MEDEVAC). You will need operational communications equipment, MEDEVAC request format, and unit signal operation instructions (SOI). Some iterations of this task should be performed in MOPP.

Some iterations of this task should be performed in MOPP.

Standards: Transmit a 9-Line MEDEVAC request, providing all necessary information as quickly as possible. Transmit, as a minimum, line numbers 1 through 5 during the initial contact with the evacuation unit. Transmit lines 6 through 9 while the aircraft or vehicle is enroute, if not included during the initial contact. IAW ATP 4-02.2, Medical Evacuation.

Special Condition: None

Special Standards: None

Special Equipment:

Safety Level: Low

MOPP: Sometimes

Task Statements

Cue: None

Note: None

References:
Required:
Related:

Performance Steps

1. Collect all applicable information needed for the MEDEVAC request.

a. Determine the grid coordinates for the pickup site. (See STP 21-1-SMCT, task 071-COM-1002.)

b. Obtain radio frequency, call sign, and suffix.

c. Obtain the number of patients and precedence.

d. Determine the type of special equipment required.

e. Determine the number and type (litter or ambulatory) of patients.

f. Determine the security of the pickup site.

g. Determine how the pickup site will be marked.

h. Determine patient nationality and status.

i. Obtain pickup site chemical, biological, radiological, and nuclear (CBRN) contamination information normally obtained from the senior person or medic.

Note: CBRN line 9 information is only included when contamination exists.

2. Record the gathered MEDEVAC information using the authorized brevity codes. (See table 081-COM-0101-1 and 081-COM-0101-2.)

Note: Unless the MEDEVAC information is transmitted over secure communication systems, it must be encrypted, except as noted in step 3b(1).

LINE	ITEM	EXPLANATION	WHERE/HOW OBTAINED	WHO NORMALLY PROVIDES	REASON
1	Location of pickup site	Encrypt the grid coordinates of the pickup site. When using the DRYAD Numeral Cipher, the same "SET" line will be used to encrypt the grid zone letters and the coordinates. To preclude misunderstanding, a statement is made that grid zone letters are included in the message (unless unit SOP specifies its use at all times).	From map	Unit leader(s)	Required so evacuation vehicle knows where to pick up patient. Also, so that the unit coordinating the evacuation mission can plan the route for the evacuation vehicle (if the evacuation vehicle must pick up from more than one location).
2	Radio frequency, call sign, and suffix	Encrypt the frequency of the radio at the pickup site, not a relay frequency. The call sign (and suffix if used) of person to be contacted at the pickup site may be transmitted in the clear.	From SOI	RTO	Required so that evacuation vehicle can contact requesting unit while en route (obtain additional information or change in situation or directions).
3	Number of patients by precedence	Report only applicable information and encrypt the brevity codes. A - URGENT B - URGENT-SURG C - PRIORITY D - ROUTINE E - CONVENIENCE If two or more categories must be reported in the same request, insert the word "BREAK" between each category.	From evaluation of patient(s)	Medic or senior person present	Required by unit controlling vehicles to assist in prioritizing missions.
4	Special equipment required	Encrypt the applicable brevity codes. A - None B - Hoist C - Extraction equipment D - Ventilator	From evaluation of patient/situation	Medic or senior person present	Required so that the equipment can be placed on board the evacuation vehicle prior to the start of the mission.
5	Number of patients by type	Report only applicable information and encrypt the brevity code. If requesting medical evacuation for both types, insert the word "BREAK" between the litter entry and ambulatory entry. L + # of patients - Litter A + # of patients - Ambulatory (sitting)	From evaluation of patient(s)	Medic or senior person present	Required so that the appropriate number of evacuation vehicles may be dispatched to the pickup site. They should be configured to carry the patients requiring evacuation.
6	Security of pickup site (wartime)	N - No enemy troops in area P - Possibly enemy troops in area (approach with caution) E - Enemy troops in area (approach with caution) X - Enemy troops in area (armed escort required)	From evaluation of situation	Unit leader	Required to assist the evacuation crew in assessing the situation and determining if assistance is required. More definitive guidance can be furnished the evacuation vehicle while it is en route (specific location of enemy to assist an aircraft in planning its approach).

Table 081-COM-0101-1

LINE	ITEM	EXPLANATION	WHERE/HOW OBTAINED	WHO NORMALLY PROVIDES	REASON
6	Number and type of wound, injury, or illness (peacetime)	Specific information regarding patient wounds by type (gunshot or shrapnel). Report serious bleeding, along with patient's blood type, if known.	From evaluation of patient(s)	Medic or senior person present	Required to assist evacuation personnel in determining treatment and special equipment needed.
7	Method of marking pickup site	Encrypt the brevity codes. A - Panels B - Pyrotechnic signal C - Smoke signal D - None E - Other	Based on situation and availability of materials	Medic or senior person present	Required to assist the evacuation crew in identifying the specific location of the pickup. Note that the color of the panels or smoke should not be transmitted until the evacuation vehicle contacts the unit (just prior to its arrival). For security, the crew should identify the color and the unit verifies it.
8	Patient nationality and status	The number of patients in each category need not be transmitted. Encrypt only the applicable brevity codes. A - US military B - US citizen C - Non-US military D - Non-US citizen E - Enemy prisoner of war (EPW)	From evaluation of patient(s)	Medic or senior person present	Required to assist in planning for destination facilities and need for guards. Unit requesting support should ensure that there is an English-speaking representative at the pickup site.
9	CBRN contamination (wartime)	Include this line only when applicable. Encrypt the applicable brevity codes. C - Chemical B - Biological R - Radiological N - Nuclear	From situation	Medic or senior person present	Required to assist in planning for the mission evacuation vehicle will accomplish the mission and when it will be accomplished).
9	Terrain description (peacetime)	Include details of terrain features in and around proposed landing site. If possible, describe relationship of site to prominent terrain feature (lake, mountain, tower).	From area survey	Personnel present	Required to allow evacuation personnel to assess route/avenue of approach into area. Of particular importance if hoist operation is required.

Table 081-COM-0101-2

a. Location of the pickup site (line 1).

b. Radio frequency, call sign, and suffix (line 2).

c. Numbers of patients by precedence (line 3).

(1) Encrypt this information using the following brevity codes: A=Urgent. B= Urgent Surgical. C= Priority. D=Routine. E= Convenience.

(2) If 2 or more categories are reported in same request, insert the word "break" between each category.

d. Special equipment required (line 4). Encrypt this information using the following brevity codes: A= None. B= Hoist. C= Extraction Equipment. D= Ventilator.

e. Number of patients by type (line 5). Encrypt this information using the following brevity codes: L+#: Number of litter patients. A+#: Number of ambulatory patients (able to walk or can walk with assistance).
Note: If requesting MEDEVAC for both types, insert the word "break" between the litter entry and the ambulatory entry.

f. Security of the pickup site (line 6- wartime). Encrypt this information using the following brevity codes: N= No enemy troops in area. P= Possibly enemy troops in area, approach with caution. E= Enemy troops in area, approach with caution. X= Enemy troops in area, armed escort required.

g. Number and type of wound, injury or illness (line 6- peacetime)

h. Method of marking the pickup site (line 7). Encrypt this information using the following brevity codes: A= Panels. B= Pyrotechnic signal. C= Smoke signal. D= None. E= Other.

i. Patient nationality and status (line 8). Encrypt this information using the following brevity codes: A= US Military. B=US Civilian. C= Non-US Military. D= Non-US Civilian. E= Enemy prisoner (EPW).

j. CBRN contamination (line 9). Encrypt this information using the following brevity codes: N= Nuclear or radiological. B= Biological. C= Chemical.

k. Terrain Description (line 9 - peacetime)

3. Transmit the MEDEVAC request. (See STP 21-1-SMCT, task 113-COM-1022.)

Note: Transmission may vary depending on individual experience level and situation.

a. Contact the unit that controls the evacuation assets.

(1) Make proper contact with the intended receiver. Use effective call sign and frequency assignments from the SOI.

(2) Give the following in the clear "I HAVE A MEDEVAC REQUEST;" wait one to three seconds for a response. If no response, repeat the statement.

b. Transmit the MEDEVAC information in the proper sequence.

(1) State all line item numbers in clear text. The call sign and suffix (if needed) in line 2 may be transmitted in the clear.text.

Note: Line numbers 1 through 5 must always be transmitted during the initial contact with the evacuation unit. Lines 6 through 9 may be transmitted while the aircraft or vehicle is en route.

(2) Follow the procedure provided in the explanation column of the MEDEVAC request format to transmit other required information. (See tables 081-COM-0101-1 and 081-COM-0101-2.)

(3) Pronounce letters and numbers according to appropriate radiotelephone procedures.

(4) End the transmission by stating "OVER."

(5) Keep the radio on and listen for additional instructions or contact from the evacuation unit.

4. Keep the radio on and listen for additional instructions or contact from the evacuation unit.

Evaluation Preparation:

Setup: For evaluation of this task, create a scenario and provide the Soldier information for the request as the Soldier requests it. You or an assistant will act as the radio contact at the evacuation unit during "transmission" of the request. Give a copy of the MEDEVAC request format to the Soldier.

Brief Soldier: Tell the Soldier to prepare and transmit a MEDEVAC request. State that the communication net is secure.

Performance Measures	GO	NO GO
1 Collected all information needed for the MEDEVAC request line items 1 through 9.	___	___

Performance Measures	GO	NO GO
Note: Wartime procedures for line items 6 and 9 will be used.		
2 Recorded the information using the authorized brevity codes.	_____	_____
3 Transmitted the MEDEVAC request as quickly as possible, following appropriate radiotelephone procedures.	_____	_____
4 Kept the radio on, listening for additional instruction or contact from the evacuation unit.	_____	_____

Evaluation Guidance: Score the Soldier GO if all steps are passed. Score the Soldier NO-GO if any step is failed. If the Soldier fails any step,show what was done wrong and how to do it correctly.

Environment: Environmental protection is not just the law but the right thing to do. It is a continual process and starts with deliberate planning. Always be alert to ways to protect our environment during training and missions. In doing so, you will contribute to the sustainment of our training resources while protecting people and the environment from harmful effects. Refer to ATP 3-34.5 Environmental Considerations and GTA 05-08-002 ENVIRONMENTAL-RELATED RISK ASSESSMENT. Environmental protection is not just the law but the right thing to do. It is a continual process and starts with deliberate planning. Always be alert to ways to protect our environment during training and missions. In doing so, you will contribute to the sustainment of our training resources while protecting people and the environment from harmful effects. Refer to ATP 3-34.5 Environmental Considerations and GTA 05-08-002 ENVIRONMENTAL-RELATED RISK ASSESSMENT.

Safety: In a training environment, leaders must perform a risk assessment in accordance with ATP 5-19, Risk Management. Leaders will complete a DD Form 2977 DELIBERATE RISK ASSESSMENT WORKSHEET during the planning and completion of each task and sub-task by assessing mission, enemy, terrain and weather, troops and support available-time available and civil considerations, (METT-TC). *Note:* During MOPP training, leaders must ensure personnel are monitored for potential heat injury. Local policies and procedures must be followed during times of increased heat category in order to avoid heat related injury. Consider the MOPP work/rest cycles and water replacement guidelines IAW TM 3-11.32 Multi-Service Reference for Chemical, Biological, Radiological, And Nuclear Warning and Reporting and Hazard Prediction Procedures.

References:
Required: ATP 4-02.2, ATP 4-25.13, STP 21-1-SMCT
Related: ATP 6-02.53

171-COM-4079

Send a Situation Report (SITREP)

Conditions: You are an element leader with an operation order (OPORD) or fragmentary order (FRAGO), map, overlay or sketch map with graphic control measures, and an operational vehicle. You may be digitally equipped. Your current situation requires you to send a SITREP. Some iterations of this task should be performed in MOPP 4.

Standards: Prepare a SITREP in standard format and send to the next higher element. Maintain situational awareness (SA).

Special Condition: None

Special Standards: None

Special Equipment:

Task Statements

Cue: None

*Note:*The operational environment must be considered at all times during this task. All Army elements must be prepared to enter any environment and perform their missions while simultaneously dealing with a wide range of unexpected threats and other influences. Units must be ready to counter these threats and influences and, at the same time, be prepared to deal with various third-party actors, such as international humanitarian relief agencies, news media, refugees, and civilians on the battlefield. These groups may or may not be hostile to us, but they can potentially affect the unit's ability to accomplish its mission.

Note: Units equipped with digital communication systems will use these systems to maximize information management, maintain SA, and minimize electronic signature.

1. Prepare a SITREP in standard format.

Note: The SITREP is used to report any change since the last report, to request resupply, and to report the current location of the element; only lines or parts of lines that contain new information will be sent. It may require additional follow-up reports.

Note: Timely and accurate reporting of friendly elements locations, obstacles and contacts are essential to maintaining SA and the reduction of potential fratricide incidents.

    a. Line 1: Date and Time Group (DTG)-Report date and time the report is being submitted.

Note: Date is the date that the report is being submitted. Time is the local time or zulu time that the report is being initiated.

    b. Line 2: Unit-Identify the unit making the report.

    c. Line 3: From-Report the time that the operational situation started or will start.

    d. Line 4: Until-Report the time that the operational situation ends or will end.

    e. Line 5: Map-Give a minimum six digit grid of the squad or team current location.

    f. Line 6: Enemy-Report enemy activity.

        (1) Nationality.

        (2) Location.

        (3) Mission.

        (4) Time of Sighting.

    g. Line 7: Nonhostile-Report nonhostile activity.

    h. Line 8: Own-Report activities of own forces.

        (1) Changes in location of units and/or formations.

        (2) Activities of forces not attached to originating unit.

2. Send the SITREP to the next higher element.

3. Maintain SA.

Evaluation Preparation:

Setup: Provide the Soldier with the equipment and or materials described in the conditions statement.

Brief Soldier: Tell the Soldier what is expected of him by reviewing the task standards. Stress to the Soldier the importance of observing all cautions, warnings, and dangers to avoid injury to personnel and, if applicable, damage to equipment.

Performance Measures	GO	NO GO
1 Prepared the SITREP in standard format.	____	____

Performance Measures	GO	NO GO
2 Sent the SITREP.	_____	_____
3 Maintained SA.	_____	_____

Evaluation Guidance: Score the Soldier GO if all performance measures are passed. Score the Soldier NO-GO if any performance measure is failed. If the Soldier scores a NO-GO, show the Soldier what was done wrong and how to do it correctly.

References:
Required: FM 6-99
Related:

171-COM-4080

Send a Spot Report (SPOTREP)

Conditions: You are an element leader with in an operation environment. You may be digitally equipped. Your current situation requires you to send a Spot Report (SPOTREP). Some iterations of this task should be performed in MOPP 4.

Standards: Prepare a Spot Report (SPOTREP) in standard format and send to the next higher element.

Special Condition: None

Safety Risk: Low

MOPP 4: Sometimes

Task Statements

Cue: None

Note: The SPOTREP is used to report timely intelligence or status regarding events that could have an immediate and significant effect on current and future operations. This is the initial means for reporting troops in contact and event information. Several lines of the SPOTREP provide sub-categories that structure reported data. Some lines may be omitted in an emergency. For example, the

SPOTREP could provide only the reporting unit, event DTG, location, and activity. The format of a SPOTREP may also change based on unit's standing operating procedures (SOP).

If equipped with Force XXI Battle Command Brigade-and-Below (FBCB2), the FBCB2 operator must update observed enemy force locations, neutral organizations, civilians and other battlefield hazards.

Performance Steps

1. Prepare SPOTREP.
   a. LINE 1 – date time group (DTG) of report submission.
   b. LINE 2 – reporting unit (Unit Making Report).

Note: After the unit designation, the method of observation must be indicated: unaided, binoculars, infrared, thermal, night vision device (NVD), unmanned aircraft system (UAS), or other. Follow with narrative if needed.

   c. LINE 3 – size of detected element.
      (1) Persons: Military, Civilian.
      (2) Vehicles: Military, Civilian.
      (3) Equipment: Military, Civilian
   d. LINE 4 - activity of detected element at DTG of report.

Note: The activity type or types must be indicated and an amplifying sub-type if Applicable. If necessary add a narrative to clarify, describe, or explain the type of activity.

      (1) Attacking (direction from).
         (a) Air defense artillery (ADA) (engaging).
         (b) Aircraft (engaging) (rotary wing [RW], fixed wing [FW]).
         (c) Ambush (IED [exploded], IED [unexploded], sniper, anti-armor, other).
         (d) Indirect fire (point of impact, point of origin).
         (e) Chemical, biological, radiological or nuclear (CBRN).
      (2) Defending (direction from).
      (3) Moving (direction from).
      (4) Stationary.
      (5) Cache.
      (6) Civilian (criminal acts, unrest, infrastructure damage).
      (7) Personnel recovery (isolating event, observed signal).
      (8) Other (give name and description).
   e. LINE 5 - location (universal transverse mercator (UTM) or grid coordinate with military grid reference system (MGRS) grid zone designator of detected element activity or event observed).
   f. LINE 6 - unit (detected element unit, organization, or facility).

Note: The type of unit, organization, or facility detected should be identified. If it cannot be clearly identified is should be described in as much detail as possible to include; the type uniform, vehicle markings, and other identifying information.

      (1) Conventional.
      (2) Irregular.

    (3) Coalition.

    (4) Host nation.

    (5) Nongovernmental organization (NGO).

    (6) Civilian.

    (7) Facility.

  g. LINE 7 – time (DTG of observation).

  h. LINE 8 – equipment (equipment of element observed).

Note: The equipment type or types, and amplifying sub-type should be identified, if applicable. A narrative can be added if necessary to clarify, describe, or explain the type of equipment. The nomenclature, type, and quantity of all equipment observed should be provided, if known. If equipment cannot be clearly identified it should be describe in as much detail as possible

    (1) ADA (missile (man-portable air defense system [MANPADS]), missile (other), gun).

    (2) Arty (gun (self-propelled [SP]), gun (towed), missile or rocket, mortar).

    (3) Armored track vehicle (tank, armored personnel carrier [APC], command and control [C2], engineer, transport, other).

    (4) Armored wheel vehicle (gun, APC, C2, engineer, transport, other).

    (5) Wheel vehicle (gun, C2, engineer, transport, other).

    (6) INF weapon (WPN) (anti-armor missile, anti-armor gun, rocket-propelled grenade [RPG], heavy [HVY] machine gun [MG], grenade launcher [GL], small arms, other).

    (7) Aircraft (RW (attack helicopter [AH]), RW (utility helicopter [UH]), RW (observation helicopter), FW (atk), FW (trans), UAS, other).

    (8) Mine or IED (buried, surface, vehicle-borne improvised explosive device [VBIED], person-borne improvised explosive device [PBIED], other).

    (9) CBRN.

    (10) Supplies (class III, class V, other).

    (11) Civilian.

    (12) Other.

  i. LINE 9 – assessment (apparent reason for or purpose of the activity observed, and apparent threats to or opportunities for friendly forces).

  j. LINE 10 –narrative (free text for clarifying report).

Note: The narrative should describe the actions taken related to the detected activity: attack, withdraw, continue to observe, or other. When feasible, the narrative should also state potential for subsequent reports such as air support request, battle damage assessment (BDA) report, call for fire, casualty report, explosive ordinance disposal (EOD) support, medical evacuation (MEDEVAC) or other reports.

  k. LINE 11 – authentication (report authentication) per SOP.

  2. Send SPOTREP to next higher element.

Note: The unit SOP may have additional guidance on who receives the SPOTREP.

Evaluation Preparation:

Setup: Provide the Soldier with the equipment and or materials described in the conditions statement.

Brief Soldier: Tell the Soldier what is expected of him by reviewing the task standards. Stress to the Soldier the importance of observing all cautions, warnings, and dangers to avoid injury to personnel and, if applicable, damage to equipment.

Performance Measures	GO	NO GO
1 Prepared the SPOTREP.	_____	_____
2 Sent the SPOTREP to higher headquarters.	_____	_____

Evaluation Guidance: Score the Soldier GO if all performance measures are passed. Score the Soldier NO-GO if any performance measure is failed. If the Soldier scores a NO-GO, show the Soldier what was done wrong and how to do it correctly.

References:
Required: FM 6-99
Related:

071-COM-0608

Use Visual Signaling Techniques

Conditions: You are a member of a mounted or dismounted platoon in a field environment and must use visual signals to communicate. Some iterations of this task should be performed in MOPP 4.

Standards: Communicate with other Soldiers and vehicle crews using visual signaling techniques.

Special Condition: None

Safety Risk: Low

MOPP 4: Sometimes

Task Statements

Cue: None

Note: Visual signals are any means of communication that require sight and can be used to transmit planned messages rapidly over short distances. This includes the devices and means used for the recognition and identification of friendly forces. The most common types of visual signals are arm-and-hand, flag, pyrotechnic, and ground-to-air. However, Soldiers are not limited to the types of signals discussed and may use what is available. Chemical light sticks, flashlights, and other items can be used, provided their use is standardized within a unit and understood by Soldiers and units working in the area. The only limit is the Soldier's initiative and imagination. Visual signals have certain limitations: (1) range and reliability of visual communications are significantly reduced during poor visibility and when terrain restricts observation; (2) may be misunderstood; and (3) vulnerable to enemy interception and may be used for deception. Leaders of mounted units use arm-and-hand signals to control individual vehicles and platoon movement. When distances between vehicles increase, flags can be used as an extension of the arms to give the signals.

Performance Steps
    1. Use visual signals for combat formations
        a. Disperse (Figure 071-326-0608-2)
            (1) Extend the arm horizontally.
            (2) Wave the arm and hand to the front, left, right, and rear
            (3) Point toward the direction of each movement.

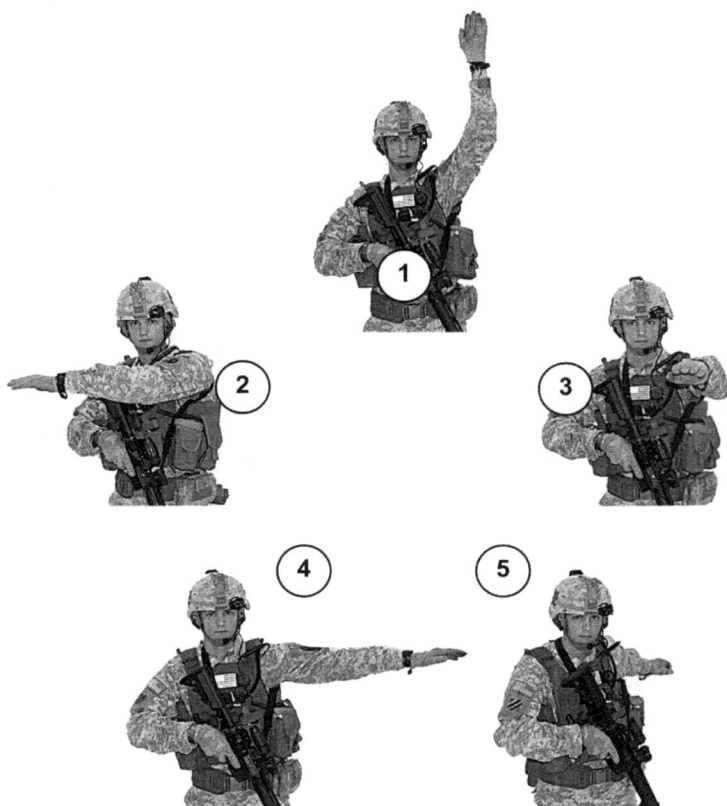

Figure 071-COM-0608-1. Disperse.

b. Assemble or Rally (Figure 071-326-0608-2).
Note: The assemble and rally signal is normally followed by pointing to the assembly or rally site.

    (1) Raise arm vertically overhead.
    (2) Turn palm to the front.
    (3) Wave in large horizontal circles.

Figure 071-COM-0608-2.
Assemble or Rally.

c. Join me, Follow me, or Come forward (Figure 071-COM-0608-3).
  (1) Point toward person(s) or unit.
  (2) Beckon by holding the arm horizontally to the front with palm up.
  (3) Motion toward the body.

Figure 071-COM-0608-3.
Join me, Follow me, or come forward.

d. Increase speed, Double time, or Rush (Figure 071-COM-0608-4).

(1) Raise the fist to the shoulder.

(2) Thrust the fist upward to the full extent of the arm and back to shoulder level.

(3) Continue rapidly several times.

Figure 071-COM-0608-4.
Increase speed, Double time, or Rush.

e. Quick time (Figure 071-COM-0608-5).

Note: This is the same signal as SLOW DOWN when directing vehicles. The difference in meaning must be understood from the context in which they are used.

(1) Extend the arm horizontally sideward.

(2) Turn palm to the front.

(3) Wave the arm slightly downward several times, keeping the arm straight.

(4) Keep arm at shoulder level.

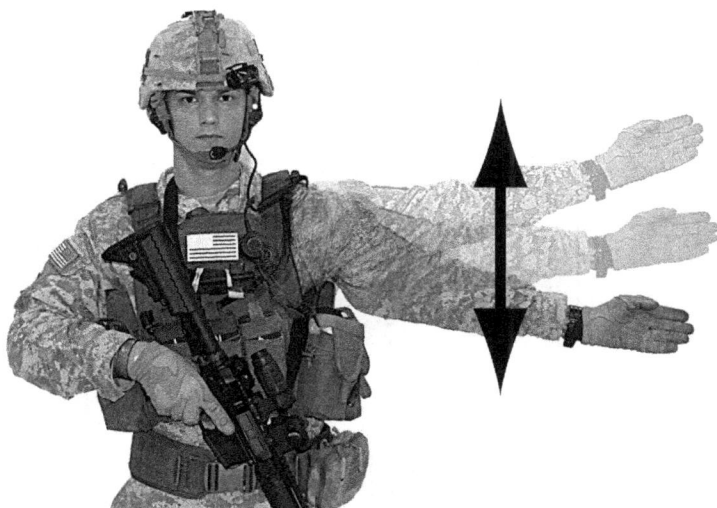

**Figure 071-COM-0608-5.
Quick Time.**

f. Enemy in sight (Figure 071-COM-0608-6).
(1) Hold the rifle in the ready position at shoulder level.
(2) Point the rifle in the direction of the enemy.

**Figure 071-COM-0608-6.
Enemy in sight.**

g. Wedge (Figure 071-COM-0608-7).
    (1) Extend the arms downward to the side.
    (2) Turn the palms to the front.
    (3) Place your arms at a 45-degree angle below horizontal.

Figure 071-COM-0608-7.
Wedge.

h. Vee (Figure 071-COM-0608-8).
   (1) Raise the arms.
   (2) Extend the arms 45-degrees above the horizontal.

Figure 071-COM-0608-8.
Vee.

i. Line (Figure 071-COM-0608-9).
    (1) Extend the arms.
    (2) Turn palms downward parallel to the ground.

Figure 071-COM-0608- 9.
Line.

j. Coil (Figure 071-COM-0608-10
    (1) Raise one arm above the head.
    (2) Rotate it in a small circle.

Figure 071-COM-0608-10.
Coil.

k. Staggered Column (Figures 11).
    (1) Extend the arms so that upper arms are parallel to the ground.
    (2) Make sure the forearms are perpendicular.
    (3) Raise the arms so they are fully extended above the head.

Figure 071-COM-0608-11.
Staggered Column.

2. Use visual signals for battle drills.
Note: Drills are a rapid, reflexive response executed by a small unit. These signals are used to initiate drills.
    a. Contact left (Figure 071-COM-0608-12).
       (1) Extend the left arm parallel to the ground.
       (2) Bend the arm until the forearm is perpendicular.
       (3) Repeat.

Figure 071-COM-0608-12.
Contact left.

b. Contact right (Figure 071-COM-0608-13).
    (1) Extend the both arms parallel to the ground.
    (2) Bend the arm until the forearm is perpendicular.
    (3) Repeat.

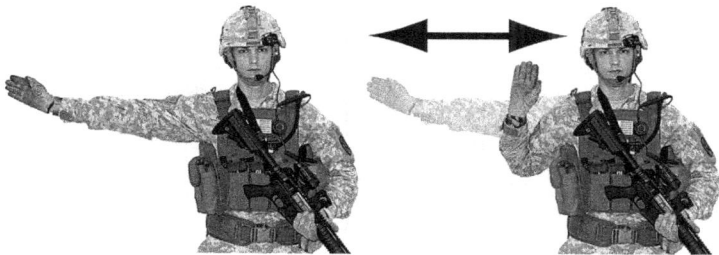

Figure 071-COM-0608-13.
Contact right.

c. Action left (Figure 071-COM-0608-14).
    (1) Extend the both arms parallel to the ground.
    (2) Raise the right arm until it is overhead.
    (3) Repeat.

Figure 071-COM-0608-14.
Action left.

d. Action right (Figure 071-COM-0608-15).
    (1) Extend the both arms parallel to the ground
    (2) Raise the left arm until it is overhead.
    (3) Repeat.

Figure 071-COM-0608-15.
Action right.

e. Air attack (Figure 071-COM-0608-16).
    (1) Bend the arms with forearms at a 45-degree angle.
    (2) The forearms should cross.
    (3) Repeat.

Figure 071-COM-0608-16.
Air Attack

    f. Nuclear, Biological, Chemical attack (Figure 071-COM-0608-17).
      (1) Extend the arms and fists.
      (2) Bend the arms to the shoulders.
      (3) Repeat.

Figure 071-COM-0608-17
Nuclear, Biological, Chemical attack.

3. Use visual signals for patrolling.

Note: Patrolling is conducted by many type units. Infantry units patrol in order to conduct combat operations. Other units patrol for reconnaissance and security. Successful patrols require clearly understood communication signals among members of a patrol.

a. Map check (Figure 071-COM-0608-18).

(1) Place one hand on top of the other.

(2) Point at the palm of one hand with the index finger of the other hand.

**Figure 071-COM-0608-18.
Map check.**

b. Pace count (Figure 071-COM-0608-19).
    (1) Bend the knee so that the heel can be tapped on.
    (2) Tap the heel of the boot repeatedly with the open hand.

Figure 071-COM-0608-19.
Pace Count.

c. Head count (Figure 071-COM-0608-20).
    (1) Raise one arm behind the head.
    (2) Tap the back of the helmet repeatedly with an open hand.

Figure 071-COM-0608-20.
Head count.

d. Danger area (Figure 071-COM-0608-21).

Note: This movement is the same as stop engine when directing a driver. The difference in meaning is understood from the context in which it is used.

(1) Raise the right hand up until it is level with the throat.

(2) Draw the right hand, palm down in a throat-cutting motion from left to right across the neck.

**Figure 071-COM-0608-21.
Danger Area.**

e. Freeze or halt (Figure 071-COM-0608-22).

(1) Make a fist with the right hand.

(1) Raise the fist to head level.

**Figure 071-COM-0608-22.
Freeze or halt.**

4. Use visual signals to control vehicle drivers.

Note: Flashlights or chemical lights are used at night to direct vehicles. Flashlights with blue filters and chemical lights will have less effect on a Soldier's night vision.

a. Start engine or prepare to move.

(1) Day: Simulate cranking of the engine by moving the arm, with the fist, in a circular motion at waist level (Figure 071-COM-0608-23).

Figure 071-COM-0608-23.
Start engine or prepare to move.

(2) Night: Move a light in a horizontal figure 8 in a vertical plane in front of the body (Figure 071-COM-0608-24).

Figure 071-COM-0608-24.
Start engine, or prepare to move (night).

b. Halt or stop.
(1) Day (Figure 071-COM-0608-25).
(a) Raise the hand upward to the full extent of the arm, with palm to the front.
(b) Hold that position until the signal is understood

**Figure 071-COM-0608-25.
Halt or stop.**

(2) Night (Figure 071-COM-0608-26).

(a) Move a light horizontally back and forth several times across the path of approaching traffic to stop vehicles.

(b) Use the same signal to stop engines.

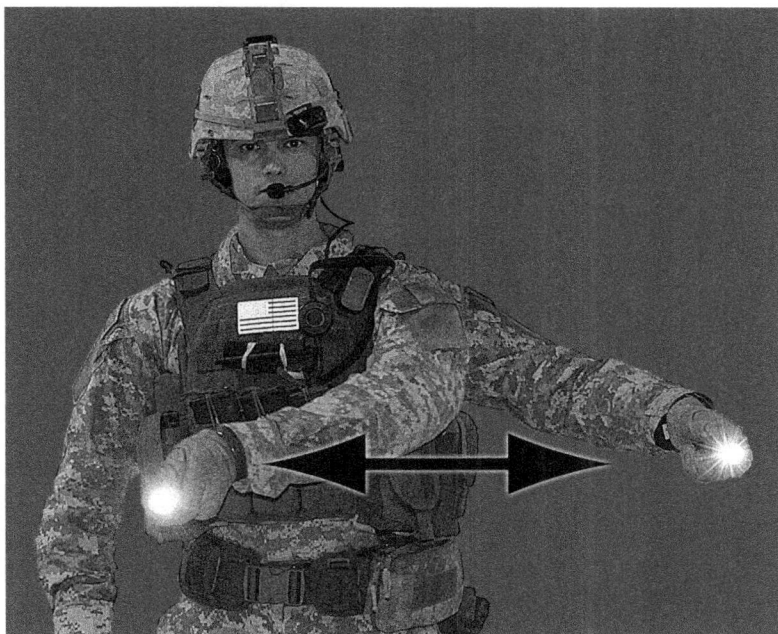

Figure 071-COM-0608-26.
Halt or Stop (night).

b. Left turn.

(1) Day (Figure 071-COM-0608-27).

(a) Extend the right arm horizontally to the side

(b) Turn palm toward vehicle with fingers extended in the direction of travel.

Figure 071-COM-0608-27.
Left turn.

(2) Night (Figure 071-COM-0608-28).
    (a) Bend the right arm at the elbow parallel to the ground.
    (b) Rotate a light to describe a 12 to 18 inch circle to the right.

Figure 071-COM-0608-28.
Left turn (night).

    d. Right Turn.
        (1) Day (Figure 071-COM-0608-29).
           (a) Extend the left arm horizontally to the side.
           (b) Turn palm toward vehicle with fingers extended in the direction
of travel.

**Figure 071-COM-0608-29.
Right turn.**

(1) Night (Figure 071-COM-0608-30).
   (a) Bend the right arm at the elbow parallel to the ground.
   (b) Rotate a light to describe a 12 to 18 inch circle to the left.

Figure 071-COM-0608-30.
Right turn (night).

e. Move forward.
    (1). Day. (Figure 071-COM-0608-31).
        (a) Face the vehicle.
        (b) Raise the hands to shoulder level with palms facing the chest.
        (c) Move the hands and forearms backward and forward.

Figure 071-COM-0608-31.
Move forward.

(1). Night (Figure 071-COM-0608-32).

    (a) Face the vehicle.

    (b) Hold a light at shoulder level.

    (c) Move the hands and forearms backward and forward.

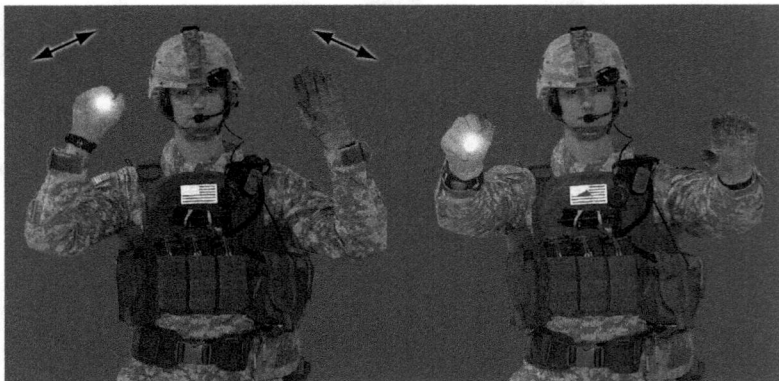

Figure 071-COM-0608-32.
Move forward (night).

f. Move in reverse.

    (1). Day (Figure 071-COM-0608-33).

        (a) Face the vehicle.

        (b) Raise the hands to shoulder level with palms facing the vehicle.

        (c) Move the hands and forearms backward and forward.

Figure 071-COM-0608-33
Move in Reverse.

(2). Night (Figure 071-COM-0608-34).
   (a) Hold a light at shoulder level.
   (b) Blink it several times toward the vehicle.

Figure 071-COM-0608-34.
Move in reverse (night).

g. Stop engine
   (1). Day (Figure 071-COM-0608-35).
      (a) Extend the arm parallel to the ground with hand open.
      (b) Move the arm across the body in a throat-cutting motion.

Figure 071-COM-0608-35.
Stop engine.

(2). Night (Figure 071-COM-0608-36).
   (a) Extend the arm parallel to the ground with hand open.

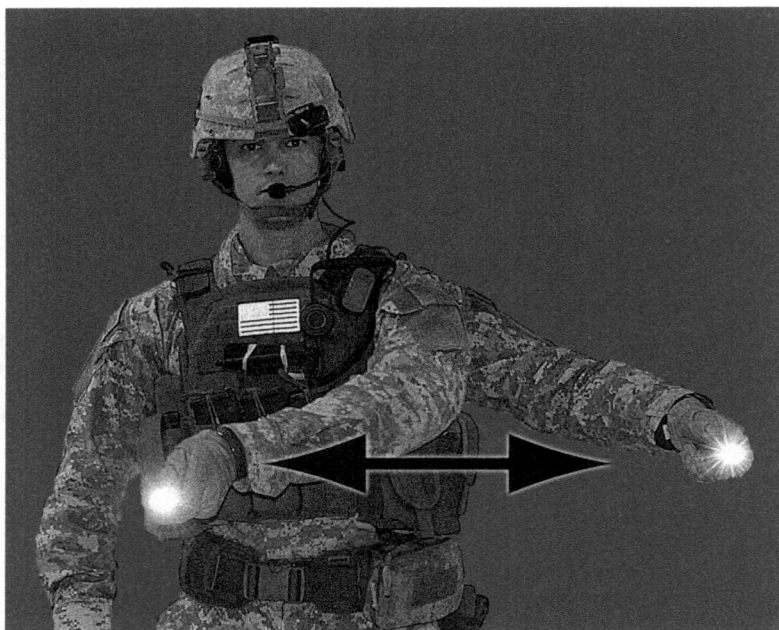

Figure 071-COM-0608-36.
Stop engine (night).

   b. Use the same signal to halt or stop vehicle.
(Asterisks indicates a leader performance step.)

Evaluation Preparation:

Setup: Provide the Soldier with the equipment and or materials described in the conditions statement.

Brief Soldier: Tell the Soldier what is expected of him by reviewing the task standards. Stress to the Soldier the importance of observing all cautions, warnings, and dangers to avoid injury to personnel and, if applicable, damage to equipment.

Performance Measures	GO	NO GO
1 Used visual signals for combat formations.	____	____
2 Used visual signals for battle drills.	____	____
3 Used visual signals for patrolling.	____	____
4 Used visual signals to control vehicle drivers.	____	____

Evaluation Guidance: Score the Soldier GO if all performance measures are passed. Score the Soldier NO-GO if any performance measure is failed. If the Soldier scores a NO-GO, show him what was done wrong and how to do it correctly.

References
Required: FM 21-60
Related:

Subject Area 5: Survive

031-COM-1010

Maintain Your Assigned Protective Mask

WARNING
READ AND ADHERE TO ALL SAFETY NOTES IN YOUR MASK'S OPERATOR'S TM PRIOR TO BEGINNING MASK MAINTENANCE.

Conditions: You are in a field or garrison environment given your assigned protective mask (with authorized accessories and components), cleaning Materials in accordance with (IAW) the applicable operator technical manual (TM), a preventive maintenance checks and services (PMCS) DA Form 5988-E Equipment Maintenance and Inspection Worksheet (EGA) or DA Form 2404 Equipment Inspection and Maintenance Worksheet, and mask replacement parts. This task should not be trained in MOPP 4.

Standards Maintain your assigned protective mask by performing operator PMCS IAW mask TM and completing DA 5988-E Equipment Maintenance and Inspection Worksheet (EGA) or DA Form 2404 Equipment Inspection and Maintenance Worksheet IAW DA Pam 750-8.

Special Condition: None

Special Standards: None

Safety Risk: Low

MOPP 4: Never

Task Statements

Cue: You are getting ready for a mission in which the protective mask is required. You are conducting regularly scheduled equipment maintenance.

WARNING
READ AND ADHERE TO ALL SAFETY NOTES IN YOUR MASK OPERATOR'S TM PRIOR TO BEGINNING MASK MAINTENANCE.

Remarks: None

Note: None

Performance Steps

1. Inspect your protective mask and accessories according to the PMCS tables located in the mask operator TM.
   a. Identify deficiencies and shortcomings.
   b. Correct operator level deficiencies.
2. Perform operator level "light" cleaning of your assigned protective mask IAW the operator TM..
3. Record uncorrected deficiencies on DA Form 5988-E or DA Form 2404 IAW DA Pam 750-8.
4. Provide the completed DA Form 5988-E or DA Form 2404 to your supervisor.

(Asterisks indicates a leader performance step.)

Evaluation Guidance: Score the Soldier GO if all performance measures are passed (P). Score the Soldier NO-GO if any performance measure is failed (F). If the Soldier fails any performance measure, show him how to do it correctly.

Evaluation Preparation:

Setup: A good time to evaluate this task is during normal care and cleaning of the mask. Place the required equipment on a field table or another suitable surface. Simulate defects in the mask by removing components from the mask or using a defective mask not issued to the Soldier.

Brief Soldier: Tell the Soldier there is no time standard for this task on the job, but for testing purposes he must perform the task within 30 minutes. Tell him to perform operator level PMCS on the mask, clean his assigned protective mask, and replace the mask filter. Tell the Soldier that completing a DA Form 2404 Equipment Inspection and Maintenance Work Sheet IAW DA PAM 750-8 is not part of the task.

	Performance Measures	GO	NO GO
1	Inspected protective mask and accessories according to the PMCS tables located in mask operator TM.	_____	_____
	a. Identify deficiences and shortcomings.		
	b. Corrected operator level deficiences.		
2	Performed operator level "light" cleaning of your assigned protective mask IAW the operator TM..	_____	_____
3	Recorded uncorrected deficiencies on a DA Form 5988-E or DA Form 2404 IAW DA Pam 750-8.	_____	_____
4	Provided the completed DA Form 5988-E or DA Form 2404 to your supervisor.	_____	_____

Evaluation Guidance: Score the Soldier GO if all performance measures are passed (P). Score the Soldier NO-GO if any performance measure is failed (F). If the Soldier fails any performance measure, show him how to do it correctly.

References:
Required: PAM 750-8, TM 3-4240-542-13&P, DA FORM 2404, DA FORM
5988-E
Related:

031-COM-1004

Protect Yourself from Chemical and Biological (CB) Contamination Using Your Assigned Protective Mask

Conditions: You are given your assigned protective mask, a hood, and a mask carrier. Some iterations of this task should be performed in MOPP 4.

Standards Do not wear contact lenses when performing this task. Do not use masks with damaged filters because filters contain.

Safety Risk: Low

MOPP 4: Sometimes

Task Statements

Cue: 1. You realize that you are under a CB agent attack, 2. You are ordered to mask, 3. You must enter a contaminated area, 4. You hear or see a chemical alarm, or 5. You observe any other automatic masking criteria designated in your unit SOP.

WARNING

Contact lenses (soft or hard) may not be worn with CB Masks. Inadequate oxygen supply to the corneal surface, exposure to dust, dirt, and smoke or gas may cause serious vision loss or eye damage. Soldiers requiring vision correction are provided optical inserts for their protective masks by their unit medical facility.

Filters must be installed prior to donning mask. Filters must be changed out one at a time. The warfighter will be unable to breathe if both filters are removed from his/ her mask. Lack of oxygen for more than 30 seconds could lead to injury or death.

Before stowing the mask, ensure that the cheek straps are not positioned below the M61 filters. Cheek straps positioned below the filters may stretch the mask causing improper chin placement; may induce buckling in the brow region causing improper seal; or may cause the cheek straps to catch underneath the filters delaying donning times resulting in illness or death.

Performance Steps

1. Don mask assembly.
   a. Stop Breathing and close eyes.
   b. Remove helmet, put helmet between legs above knees or hold rifle between legs and place helmet on the muzzle.
Note: If helmet falls continue to mask.
   c. Take off glasses and place in helmet, if applicable.
   d. Open the mask carrier with left hand.
   e. Grasp the mask assembly with right hand, and remove it from the carrier.
   f. Place chin in the chin pocket, and press the face piece tight against face.
   g. Hold mask assembly tightly against face.
   h. Grasp the harness tab, pull overhead and down the head harness as far as possible.
Note: Ensuring the head harness is centered on the crown of the head and the temple straps are approximately parallel to the ground.
   i. Grasp the loose end of the cheek straps, one at a time, and pull until strap feels tight.

Note: Both straps should be approximately equal length when complete. Also the temple and forehead straps have already been adjusted during fitting do not tighten.
2. Clear mask assembly.

    a. Seal the outlet disk valve by placing one hand over the outlet valve cover assembly.

    b. Blow out hard to ensure that any contaminated air is forced out around the edges of the face piece.

  3. Seal mask assembly.

    a. Cover both filter inlet ports with the palms of your hands and breathe in.

    b. Ensure mask assembly collapse against the face.

    c. Resume breathing.

  4. Give the alarm.

    a. Shout, "Gas, Gas, Gas."

    b. Give the appropriate hand-and-arm signal per unit SOP.

  5. Close mask carrier.

  6. Don the mask hood, if applicable.

Note: the Soldier is using the mask in conjunction with the joint-service, lightweight integrated suit technology (JSLIST), he/she skips this step (the mask lacks a hood because it is built in on the JSLIST).Be careful when pulling on the hood because it could snag and tear on the buckles of the head harness. Be careful not to break face piece seal when pulling protective hood over your head.

    a. Place hands up under the protective hood, stretch elasticized portion and raise protective hood up and over filters.

    b. Carefully pull excess protective hood material over head, neck and shoulders.

    c. Grasp underarm straps.

    d. Bring the male end of each underarm strap and fasten to female end.

    e. Tighten underarm straps.

  7. Put on the helmet and secure gear.

Note: For combat vehicle crewman (CVC) helmet, perform the following steps: 1. Disconnect the boom microphone from the helmet, 2. Connect the mask microphone to the receptacle in the helmet, 3. Grasp the helmet next to the ear cups with the hand, and spread the helmet as far as possible, 4. Place the helmet overhead, tilting the helmet forward slightly so that the first contact when putting it on is with the forehead surface of the mask and 5. Rotate the helmet back and down over the head until it is seated in position. Make sure you don't break seal of the mask.

  8. Continue the mission.

(Asterisks indicates a leader performance step.)

Evaluation Guidance: Read the action, condition, and standard to the Soldier. Provide the Soldier with all items given in the Conditions Statement. Score the Soldier GO if all performance measures are passed (P) in sequence. Soldier must complete steps 1 through 3, in sequence, within 9 seconds. Score the Soldier NO GO if any performance measure is failed (F) or out of sequence. If the Soldier scores NO GO, show the Soldier what was done wrong and how to do it correctly.

Evaluation Preparation: Gather all items in the conditions statement so that they can be provided to the Soldier. Prepare scenarios and questions to ask the Soldier in order to illicit a response that can be evaluated against the performance measures.

Performance Measures	GO	NO GO
1 Donned the mask.	_____	_____

a. Stopped Breathing and closed eyes.

b. Removed helmet, put helmet between legs above knees or held rifle between legs and placed helmet on the muzzle.

c. Took off glasses and placed in helmet, if applicable.

d. Opened the mask carrier with left hand.

e. Grasped the mask assembly with right hand, and removed it from the carrier.

f. Placed chin in the chin pocket, and pressed the face piece tight against face.

g. Held mask assembly tightly against face.

h. Grasped the harness tab, pulled over head and down the head harness as far as possible. Note:

i. Grasped the loose end of the cheek straps, one at a time, and pulled until strap feels tight.

2 Cleared mask assembly.	_____	_____

a. Sealed the outlet disk valve by placing one hand over the outlet valve cover assembly.

b. Blew out hard and ensured that any contaminated air is forced out around the edges of the face piece.

3 Checked mask Assembly.	_____	_____

Performance Measures	GO	NO GO
a. Covered both filter inlet ports with the palms of your hands and breathed in.		
b. Ensured mask assembly collapse against the face.		
c. Resumed breathing.		
4 Gave the alarm.	____	____
a. Shouted, "Gas, Gas, Gas."		
b. Gave the appropriate hand-and-arm signal per unit SOP.		
5 Closed mask carrier.	____	____
6 Donned the mask hood, if applicable.	____	____
a. Placed hands up under the protective hood, stretched elasticized portion and raised protective hood up and over filters.		
b. Carefully pulled excess protective hood material over head, neck and shoulders.		
c. Grasped underarm straps.		
d. Brought the male end of each underarm strap and fastened to female end.		
e. Tightened underarm straps.		
7 Put on the helmet and secured gear.	____	____
8 Continued the mission.	____	____

Evaluation Guidance: Score the Soldier GO if all performance measures are passed (P). Score the Soldier NO-GO if any performance measure is failed (F). If the Soldier fails any performance measure, show him how to do it correctly.

Environment: Environmental protection is not just the law but the right thing to do. It is a continual process and starts with deliberate planning. Always be alert to ways to protect our environment during training and missions. In doing so, you will contribute to the sustainment of our training resources while protecting people and the environment from harmful effects. Refer to the current Environmental Considerations manual and the current GTA Environmental-related Risk Assessment card.

Safety: In a training environment, leaders must perform a risk assessment in accordance with current Risk Management Doctrine. Leaders will complete the current Deliberate Risk Assessment Worksheet in accordance with the TRADOC Safety Officer during the planning and completion of each task and sub-task by assessing mission, enemy, terrain and weather, troops and support available-time available and civil considerations, (METT-TC).

Note: During MOPP training, leaders must ensure personnel are monitored for potential heat injury. Local policies and procedures must be followed during times of increased heat category in order to avoid heat related injury. Consider the MOPP work/rest cycles and water replacement guidelines IAW current CBRN doctrine.

References:
Required: TM 3-11.32, TM 3-4240-542-13&P
Related:

031-COM-1007

React to Chemical or Biological (CB) Hazard/Attack

Foreign Disclosure: FD1 - This training product has been reviewed by the training developers in coordination with the USACBRNS, Foreign disclosure representative and MSCoE foreign disclosure officer. This training product can be used to instruct international military students from all approved countries without restrictions.

Conditions: You are in an area in which a chemical or biological attack is occurring. You are given your assigned protective mask and complete set of MOPP Gear (JSLIST, Gloves, and Boots), individual equipment decontamination kit, and Reactive Skin Decontamination Lotion (RSDL). You are wearing eye protection, Army Combat Helmet (ACH), Improved Outer Tactical Vest (IOTV), and Deltoid Auxiliary Protectors (DAPs). You are currently in MOPP Level 0.

Standards: React to a CB hazard/attack without becoming a casualty by: donning your protective mask within 9 seconds, starting skin decontamination yourself within 1 minute of finding contamination, assuming MOPP 4 (after decontamination)Â and decontaminating your individual equipment using the decontaminating kit as necessary.

.Special Condition: None

Special Standards: None

Safety Risk: Low

Task Statements

Cue: None

Note: None

Performance Steps

1. Protect yourself from CB contamination by using your assigned protective mask, IAW common task 031-COM-1004, within 9 seconds.
Note: The mask provides protection against conventional warfare agents. The mask provides little if any protection from toxic industrial materials (TIMs), but it provides the best available protection to enable you to evacuate the hazard area. You may be required to evacuate to a minimum safe distance of at least 300 meters upwind from the contamination (if possible) or as directed by the commander.

2. Take cover (if possible) and conduct immediate skin decontamination, IAW common task 031-COM-1006, within 1 minute.

3. Assume MOPP Level 4, IAW common task 031-COM-1005, within 8 minutes. Follow the step below when wearing ACH, IOTV, or DAPs.
   a. Remove the ACH and protective eye wear.
   b. Loosen the DAPs.

WARNING
WHEN DOFFING THE IOTV FROM THE SHOULDER. TAKE CARE NOT TO SNAG THE FILTER CANISTER AND BREAK THE SEAL OF YOUR PROTECTIVE MASK.

---

    c. Doff the IOTV by lifting the front flap and detach side plate carriers by separating hook and loop fastener tape. Lift front carrier and detach internal elastic bands at hook and loop interface. Open the medical access hook and pile closure, loosen the left shoulder adjustment strap and slide vest off the right shoulder.

    d. Assume MOPP Level 4.

    e. Don the IOTV over the right shoulder by tightening the left shoulder adjustment strap and fastening the medical access hook and pile closure. Attach internal elastic bands at hook and loop interface and close the front carrier. Attach side plate carriers and close the front flap.

    f. Secure the DAP.

    g. Don the ACH.

    4. Decontaminate your individual equipment using your individual equipment decontamination kit, IAW common task 031-COM-1006.

(Asterisks indicates a leader performance step.)

Evaluation Guidance: Score the Soldier GO if all performance measures are passed (P). Score the Soldier NO-GO if any performance measure is failed (F). If the Soldier fails any performance measure, show him how to do it correctly.

Evaluation Preparation: Setup: A good time to evaluate this task is during a field exercise when a variety of CB hazards can be simulated.
Select a site with adequate cover, and ensure that Soldiers have all items listed in the task conditions.

Brief Soldier: Give the Soldier being tested the chemical attack alarm, "GAS, GAS, GAS".

Evaluation Preparation:

Setup: A good time to evaluate this task is during a field exercise when a variety of CB hazards can be simulated. Select a site with adequate cover, and ensure that Soldiers have their assigned protective mask.

Brief Soldier: Tell the Soldier that there will be an encounter with simulated CB agents and/or a CB alarm will be given

Performance Measures	GO	NO GO
1 Protected yourself from CB contamination by using your assigned protective mask, IAW common task 031-COM-1004, within 9 seconds.	_____	_____

Performance Measures	GO	NO GO
2 Took cover (if possible) and conducted immediate skin decontamination, IAW common task 031-COM-1006, within 1 minute.	_____	_____
3 Assumed MOPP Level 4, IAW common task 031-COM-1005, within 8 minutes. Followed the step	_____	_____

a. Removed the ACH and Protective eyewear.

b. Loosened the DAPs.

c. Doffed the IOTV by lifting the front flap and detaching side plate carriers by separating hook and loop fastener tape. Lifted front carrier and detached internal elastic bands at hook and loop interface. Opened the medical access hook and pile closure, loosened the left shoulder adjustment strap and slid vest off the right shoulder.

d. Assumed MOPP Level 4.

e. Donned the IOTV over the right shoulder by tightening the left shoulder adjustment strap and fastening the medical access hook and pile closure. Attached internal elastic bands at hook and loop interface and closed the front carrier. Attached side plate carriers and closed the front flap.

f. Secured the DAPs.

g. Donned the ACH.

below when wearing ACH, IOTV, or DAPs.

4 Decontaminated your individual equipment using your individual equipment decontamination kit, IAW common task 031-COM-1006.	_____	_____

Evaluation Guidance: Score the Soldier GO if all performance measures are passed (P). Score the Soldier NO-GO if any performance measure is failed (F). If the Soldier fails any performance measure, show him how to do it correctly.

References:
Required: ATP 3-11.32, TM 10-8415-220-10, TM 3-4230-235-10, TM 3-4240-542-13&P.
Related:

Environment: Environmental protection is not just the law but the right thing to do. It is a continual process and starts with deliberate planning. Always be alert to ways to protect our environment during training and missions. In doing so, you will contribute to the sustainment of our training resources while protecting people and the environment from harmful effects. Refer to the current Environmental Considerations manual and the current GTA Environmental-related Risk Assessment card.

Safety: In a training environment, leaders must perform a risk assessment in accordance with current Risk Management Doctrine. Leaders will complete the current Deliberate Risk Assessment Worksheet in accordance with the TRADOC Safety Officer during the planning and completion of each task and sub-task by assessing mission, enemy, terrain and weather, troops and support available-time available and civil considerations, (METT-TC).

Note: During MOPP training, leaders must ensure personnel are monitored for potential heat injury. Local policies and procedures must be followed during times of increased heat category in order to avoid heat related injury. Consider the MOPP work/rest cycles and water replacement guidelines IAW current CBRN doctrine.

031-COM-1005

Protect Yourself from CBRN Injury/Contamination by Assuming MOPP Level 4

Foreign Disclosure: FD1 - This training product has been reviewed by the training developers in coordination with the USACBRNS Foreign Disclosure (FD) representative and MSCoE foreign disclosure officer. This training product can be used to instruct international military students from all approved countries without restrictions.

Conditions: In a contaminated or potentially contaminated environment, given the Joint-Service, Lightweight, Integrated Suit Technology (JSLIST), your assigned protective mask, CBRN over boots, and CBRN protective gloves. You are in MOPP level 0. This task is always performed in MOPP 4.

Standards: Protect yourself from CBRN injury or contamination by assuming MOPP level 1 thru 4 in sequence within eight minutes.

Special Condition: None

Safety Risk: Low

MOPP 4: Always

Task Statements

Special Equipment:

Cue: You learn a CBRN attack is imminent or must enter/cross an area where CBRN has been used. You are directed to assume mission-orientedprotective posture (MOPP) level 4.

Note: Complete Steps in sequence within eight minutes for a GO.

Performance Steps

1. Assume MOPP Level 1 by donning the JSLIST over garments.
Note: Complete Steps in sequence within eight minutes for a GO.
   a. Don the JSLIST over garment trousers.
     (1) Extend your toes downward, put one leg into the trousers, and pull them up. Repeat the procedure for your other leg.
     (2) Close the slide fastener, and fasten the two fly opening snaps.
     (3) Pull the suspenders over your shoulders, and fasten the snap couplers.
     (4) Adjust the suspenders to ensure that the trousers fit comfortably.
Note: The trouser length can be adjusted by raising or lowering the suspenders.
     (5) Adjust the waistband hook-and-pile fasteners for a snug fit.
   b. Don the JSLIST over garment coat.
     (1) Don the coat, and close the slide fastener up as far as your chest.
     (2) Secure the front closure hook-and-pile fasteners up as far as your chest.
     (3) Pull the bottom of the coat down over the trousers.
     (4) Pull the loop out and away from the over garment coat, and bring it forward between the legs.
     (5) Pull on the loop until the bottom of the coat fits snugly over the trousers.
2. Assume MOPP Level 2 by donning the over boots.
   a. Don the over boots over the combat boots.
   b. Adjust and secure the strap-and-buckle fasteners.
   c. Pull the trouser legs over the over boots.
   d. Secure the hook-and-pile fasteners on each ankle to fit snugly around the boot.

3. Assume MOPP Level 3 by donning chemical-protective mask IAW task 031-COM-1004.

4. Assume MOPP Level 4. Don the gloves..

    a. Push the sleeve cuffs up your arm.

    b. Put on the glove liners (inserts).

    c. Put on the gloves (black rubber).

    d. Pull the sleeve cuffs over the top of the gloves.

(Asterisks indicates a leader performance step.)

Evaluation Guidance: Read the action, condition, and standard to the Soldier. Provide the Soldier with all items given in the Conditions Statement. Complete Steps in sequence within eight minutes for a GO. Score the Soldier GO if all performance measures are passed (P) in sequence. Score the Soldier NO GO if any performance measure is failed (F) or out of sequence. If the Soldier scores NO GO, show the Soldier what was done wrong and how to do it correctly.

Evaluation Preparation: Gather all items in the conditions statement so that they can be provided to the Soldier. Prepare scenarios and questions to ask the Soldier in order to illicit a response that can be evaluated against the performance measures.

Performance Measures	**GO**	**NO GO**
1 Assumed MOPP Level 1, donned the JSLIST over gaements.	_____	_____

Donned the JSLIST over garment trousers.

(1) Extended their toes downward, put one leg into the trousers, and pulled them up. Repeated the procedure for their other leg.

(2) Closed the slide fastener, and fastened the two fly opening snaps.

(3) Pulled the suspenders over their shoulders, and fastened the snap couplers.

(4) Adjusted the suspenders to ensure that the trousers fit comfortably.

b. Donned the JSLIST over garment coat.

Performance Measures	GO	NO GO

(1) Donned the coat, and closed the slide fastener up as far as their chest.

(2) Secured the front closure hook-and-pile fasteners up as far as their chest.

(3) Pulled the bottom of the coat down over the trousers.

(4) Pulled the loop out and away from the over garment coat, and brought it forward between the legs.

(5) Pulled on the loop until the bottom of the coat fits snugly over the trousers.arments.

2   Aa. Donned the over boots over the combat boots. _____ _____

b. Adjusted and secured the strap-and-buckle fasteners.

c. Pulled the trouser legs over the over boots.

d. Secured the hook-and-pile fasteners on each ankle to fit snugly around the boot.ssumed MOPP Level 2, donned the over boots.

3   Assumed MOPP Level 3, donned chemical-protective mask IAW task 031-COM-1004.. _____ _____

4   Assumed MOPP Level 4. Don the gloves. _____ _____

a. Pushed the sleeve cuffs up arm.

b. Put on the glove liners (inserts).

Performance Measures	GO	NO GO
c. Put on the gloves (black rubber).		
d. Pulled the sleeve cuffs over the top of the gloves.		
e. Secured the hook-and-pile fastener tape snugly on each wrist.		

Evaluation Guidance: Score the Soldier GO if all performance measures are passed (P). Score the Soldier NO-GO if any performance measure is failed (F). If the Soldier fails any performance measure, show him how to do it correctly.

References:
Required: TM 3-11.32, TM 10-8415-220-10
Related:

Environment: Environmental protection is not just the law but the right thing to do. It is a continual process and starts with deliberate planning. Always be alert to ways to protect our environment during training and missions. In doing so, you will contribute to the sustainment of our training resources while protecting people and the environment from harmful effects. Refer to the current Environmental Considerations manual and the current GTA Environmental-related Risk Assessment card.

Safety: In a training environment, leaders must perform a risk assessment in accordance with current Risk Management Doctrine. Leaders will complete the current Deliberate Risk Assessment Worksheet in accordance with the TRADOC Safety Officer during the planning and completion of each task and sub-task by assessing mission, enemy, terrain and weather, troops and support available-time available and civil considerations, (METT-TC).

Note: During MOPP training, leaders must ensure personnel are monitored for potential heat injury. Local policies and procedures must be followed during times of increased heat category in order to avoid heat related injury. Consider the MOPP work/rest cycles and water replacement guidelines IAW current CBRN doctrine.

031-COM-1009

Detect Chemical Agents Using M9 Detector Paper

Foreign Disclosure: FD3 - This training product has been reviewed by the developers in coordination with the MSCoE Foreign Disclosure Officer foreign disclosure officer. This training product cannot be used to instruct international military students.

Conditions: You are in tactical environment or an area with suspected liquid chemical contamination given a roll of M9 detector paper. Some iterations of this task should be performed in MOPP 4.

Standards: Detect liquid chemical agents by attaching M9 detector paper to your MOPP gear on the correct (according to Â dominant hand)Â upper arm, wrist, and ankle with 1 inch tabs' Attach to equipment, with 1.5 inch tabs, in areas likely to be contaminated within view of the operator IAW unit SOP.

Special Condition: None.

Special Standards: None

MOPP 4 : Sometimes

Task Statements

Cue: 1. you are required to initiate passive defensive measures prior to or after a CBRN attack.
2. You are required to detect a potential liquid chemical agent.

WARNING

Always wear protective gloves when touching M9 detector paper. Do not get M9 detector paper in or near your mouth or on your skin.

CAUTION

Firing weapons lubricated with lubricating oil, semi-fluid; lubricant, small arms; or lubricant, semifluid, automatic weapons (LSA) may cause false positive responses on the olive drab (OD) detector paper.

Heat may cause detector paper to turn red and cause false readings. Keep detector paper away from hot surfaces such as vehicle hoods, weapon barrels, or cannon assemblies.

Do not attach M9 detector paper to hot, dirty, oily, or greasy surfaces because it may give false positive readings.

Do not rub or scrape detector paper across rough surfaces. Scuff marks will cause false readings.

A color blind person may see a red spot as gray or black. Spots must be checked by a person who is not color blind.

When dispenser is not in use, place in plastic storage bag to prevent contamination.

Notes: Do not check M9 detector paper with colored light because you will not see liquid chemical agent red spots.

1. Attach M9 detector paper to your MOPP gear.

Note: M9 detector paper will not detect chemical agent vapors. Paper band must not be too tight because it will tear with movement. However, it must not be too loose because it may slip down. If assistance is available, let your buddy tear off and attach your detector paper to your MOPP gear.

a. If you are left handed, place a strip of M9 detector paper around your left upper arm, right wrist, and left ankle with approximately 1 inch overlaps.

b. If you are right-handed, place a strip of M9 detector paper around your right upper arm, left wrist, and right ankle with approximately 1 inch overlaps.

Note: These are the places where a moving Soldier will most likely brush against a surface (such as undergrowth) that is contaminated with a liquid chemical agent.

2. Attach M9 paper to equipment.

a. Place M9 paper with 1.5 inch tab for easy removal where it will come into contact with contaminated objects.

b. Place M9 paper where it will be visible to the operator.

3. Check for surface liquid agent contamination by taking a piece of M9 paper and blotting the surface of equipment, ground, or vegetation around suspected area.

4. Monitor the M9 detector paper constantly for any color change.

Note: If pink, red, red-brown, red-purple, or any shade of red streaks or spots are detected assume that you have been exposed to a liquid chemical agent. Blue, yellow, green, gray, or black spots are not from a liquid chemical agent.

5. Use other types of chemical agent detectors kits (such as M8 paper or M256A2 kit) to verify the test results.

6. Notify supervisor or the results.

Evaluation Preparation:

Setup: Provide the items listed in the task condition statement. Simulate an unknown liquid chemical agent by using expedient training aids (such as brake fluid, cleaning compound, gasoline, insect repellent, or antifreeze). Place drops of the simulated agent on M9 detector paper to obtain a reading. For M8 detector paper, place the simulated agent on non   porous material (such as an entrenching tool).

Brief Soldier: Tell the Soldier that he/she will be entering an area where chemical agents have been used. Tell him/her to attach M9 detector paper to his/her MOPP gear and equipment. Tell him/her that if you observe any acts that are unsafe or that could produce a false reading you will stop the test and he/she will be scored a NO GO.

Performance Measures	GO	NO GO
1. Attached M9 detector paper to Soldiers in MOPP gear.	_____	_____

a. Placed strip of M9 detector paper around your left upper arm, right wrist, and left ankle if you wereleft-handed with approximately 1 inch overlaps.

b. Placed M9 detector paper around your right upper arm, left wrist, and right ankle if you were righthanded with approximately 1 inch overlaps.

2. Attached M9 paper to equipment.	_____	_____

a. Placed M9 detector paper with 1.5 inch tab for easy removal where it would come in contact withcontaminated objects.

b. Placed M9 detector paper where it will be seen by the operator.

3. Checked for surface liquid agent contamination by taking a piece of M9 paper and blotting the surfaceof equipment, ground, or vegetation around suspected area.	_____	_____
4. Monitored the M9 detector paper constantly for any color change.	_____	_____
5. Used other types of chemical agent detectors kits (such as M8 paper or M256A2 kit) to verify the test results.	_____	_____
6. Notified supervisor of the results.	_____	_____

Evaluation Guidance: Score the Soldier GO if all performance measures are passed (P). Score the Soldier NO-GO if any performance measure is failed (F). If the Soldier fails any performance measure, show him how to do it correctly.

References:
Required: TM 3-6665-311-10

Related:

Environment: Environmental protection is not just the law but the right thing to do. It is a continual process and starts with deliberate planning. Always be alert to ways to protect our environment during training and missions. In doing so, you will contribute to the sustainment of our training resources while protecting people and the environment from harmful effects. Refer to the current Environmental Considerations manual and the current GTA Environmental-related Risk Assessment card. Environmental considerations and hazards (Noise, Air, Water, or Land Pollution) associated with this training must be reviewed as published in the required and related references before conducting demonstration and hands-on training for this equipment. Follow procedures to dispose of waste materials per local environmental restrictions.

Safety: In a training environment, leaders must perform a risk assessment in accordance with current Risk Management Doctrine. Leaders will complete the current Deliberate Risk Assessment Worksheet in accordance with the TRADOC Safety Officer during the planning and completion of each task and sub-task by assessing mission, enemy, terrain and weather, troops and support available-time available and civil considerations, (METT-TC).

Note: During MOPP training, leaders must ensure personnel are monitored for potential heat injury. Local policies and procedures must be followed during times of increased heat category in order to avoid heat related injury. Consider the MOPP work/rest cycles and water replacement guidelines IAW current CBRN doctrine.

031-COM-1008

Identify Liquid Chemical Agents using M8 Paper

Foreign Disclosure: FD3 - This training product has been reviewed by the developers in coordination with the MSCoE Foreign Disclosure Officer foreign disclosure officer. This training product cannot be used to instruct international military students.

Conditions: You are in a tactical environment or an area with suspected liquid chemical contamination, given a booklet of M8 detector paper. This task is always performed in MOPP 4.

Standards: Identify liquid chemical agents using M8 Paper by comparing any color change on the M8 detector paper to the color chart on the inside front cover of the booklet with 100% accuracy.

Special Conditions: None
Safety Risk: Low
MOPP 4: Always

Task Statements

Cue:
   1. You are required to initiate passive CBRN defensive measures prior to or after a CBRN attack.
   2. You are required to detect a potential liquid chemical agent.

WARNING

Do not use M8 paper if you do not see colors correctly. Color combinations and comparisons are used during test. A wrong reading of results might cause agent exposure due to premature removal of protective equipment.

M8 paper that indicates positive results should be treated as contaminated. Report results, as required.

Notes: M8 detector paper reacts positively with petroleum products and ammonia. When conducting agent test at night, remove any colored lens because it may provide a false negative response.

Performance Steps

   1. Identify liquid chemical agents with M8 detector paper using the active method.
      a. Remove a sheet of M8 paper from the book (use one-half sheet if it is perforated).
Note: You may want to put the paper on the end of a stick or another object and then blot the paper on the suspected liquid agent.

CAUTION
M8 paper is subject to false positive indications caused by many substances. Do not scrub, or rub M8 paper on suspected contaminated surfaces.

---

b. Dip the paper into the suspected liquid agent or blot the suspected area to be tested with the paper.

Note: Do not touch the liquid with protective glove.

WARNING

Some decontaminants will give false positive results on the M8 detector paper. The M8 detector paper may indicate positive results if used in an area where decontaminants have been used.

---

c. Observe the M8 detector paper for a color change. Identify the contamination by comparing any color change on the M8 detector paper to the color chart on the inside front cover of the booklet.

(1) A yellow-gold color indicates the presence of a nerve (G) agent.

(2) A red-pink color indicates the presence of a blister (H) agent.

(3) A dark green color indicates the presence of a nerve (V) agent.

(4) Any other color or no color change indicates that the liquid cannot be identified using M8 detector paper.

2. Identify liquid chemical agents with M8 detector paper using the passive method.

a. Remove a sheet of M8 paper from the booklet.

b. Secure the sheet to any object in an area which would most likely receive contamination.

c. Periodically inspect the paper for color changes. Identify the contamination by comparing any color change on the M8 detector paper to the color chart on the inside front cover of the booklet.

3. Store the booklet of M8 detector paper in a manner which will prevent wetting.

4. Use other types of chemical-agent detector kits (such as the M256A2 Kit) to verify the test results.

5. Notify your supervisor of the test results.

(Asterisks indicates a leader performance step.)

Evaluation Guidance: Score the Soldier GO if all performance measures are passed (P). Score the Soldier NO-GO if any performance measure is failed (F). If the Soldier scores NO-GO, show the Soldier what was done wrong and how to do it correctly.

Evaluation Preparation: Provide the Soldier with the items listed in the task condition statement. Evaluate this task during a field training exercise or a situational training exercise.

Performance Measures	GO	NO GO
1 Identified liquid chemical agents with M8 detector paper using the active method.	____	____

a. Removed a sheet of M8 detector paper from the book.
b. Dipped the paper into the suspected liquid agent without touching the liquid with protective glove.
c. Observed the M8 detector paper for a color change. Identified the contamination by comparing any color change on the M8 detector paper to the color chart on the inside front cover of the booklet.

	GO	NO GO
2 Identified liquid chemical agents with M8 detector paper using the passive method.	____	____

a. Removed a sheet of M8 paper from the booklet.
b. Secured the sheet to an object in the area which would most likely receive contamination.
c. Periodically inspected the paper for color changes. Identified the contamination by comparing any color change on the M8 paper to the color chart on the inside front cover of the booklet.

	____	____
3. Stored the booklet of M8 detector paper.		
4. Used other types of chemical-agent detector kits (such as the M256A2 kit) to verify the test results.	____	____
5. Notified supervisor of the test results.	____	____

Evaluation Guidance: Score the Soldier GO if all performance measures are passed (P). Score the Soldier NO-GO if any performance measure is failed (F). If the Soldier fails any performance measure, show him how to do it correctly.

Required:

Environment: Environmental protection is not just the law but the right thing to do. It is a continual process and starts with deliberate planning. Always be alert to ways to protect our environment during training and missions. In doing so, you will contribute to the sustainment of our training resources while protecting people and the environment from harmful effects. Refer to the current Environmental Considerations manual and the current GTA Environmental-related Risk Assessment card. Environmental considerations and hazards (Noise, Air, Water, or Land Pollution) associated with this training must be reviewed as published in the required and related references before conducting demonstration and hands-on training for this equipment. Follow procedures to dispose of waste materials per local environmental restrictions.

Safety: In a training environment, leaders must perform a risk assessment in accordance with current Risk Management Doctrine. Leaders will complete the current Deliberate Risk Assessment Worksheet in accordance with the TRADOC Safety Officer during the planning and completion of each task and sub-task by assessing mission, enemy, terrain and weather, troops and support available-time available and civil considerations, (METT-TC).
Note: During MOPP training, leaders must ensure personnel are monitored for potential heat injury. Local policies and procedures must be followed during times of increased heat category in order to avoid heat related injury. Consider the MOPP work/rest cycles and water replacement guidelines IAW current CBRN doctrine. Everyone is responsible for safety. A thorough risk assessment must be completed prior to every mission or operation.

031-COM-1006

Decontaminate Yourself and Individual Equipment Using Chemical Decontaminating Kits

Foreign Disclosure: FD1 - This training product has been reviewed by the training developers in coordination with the MSCoE foreign disclosure officer. This training product can be used to instruct international military students from all approved countries without restrictions.

Conditions: You are in a contaminated environment in MOPP Level 2. You are given a chemical protective mask, chemical protective gloves, chemical protective overboots, a full canteen of water, a poncho, load-bearing equipment (LBE) or load-bearing vest, Interceptor Body Armor (IBA), the Improved Outer Tactical Vest (IOTV), and M295 decontaminating kitÂ and Reactive Skin Decontamination Lotion (RSDL). This task is always performed in MOPP 4.

Standards: Decontaminate yourself and your individual equipment using the chemical decontaminating kits. Start the steps to decontaminate your skin and eyes within 1 minute after contamination. and finish within 2 minutes. Decontaminate all individual equipment, in sequence, within 15 minutes after

decontaminating your skin..

Special Condition: None

Special Standards: None

Special Equipment:

Task Statements

Cue: Your skin and equipment have been exposed to chemical agents, or you have passed through a chemically contaminated area and suspect that you have contamination on your skin.

DANGER
Do Not mix RSDL with solid, undiluted high-test hypochlorite (HTH) or super tropical bleach (STB), heat and/or fire may result.

WARNING
Under no circumstances should the training RSDL be used in place of the RSDL during actual combat operations. The training lotion does not contain active ingredients.

Performance Steps

1. Don protective mask and hood. Do not pull the drawstrings. Do not fasten the shoulder straps if your hood has them.
Note: For training purposes, use the Training RSDL.

2. Seek overhead cover or use a poncho for protection against further contamination.
Note: If contamination of the eyes is suspected, stop breathing, remove mask and place in on an uncontained surface, if available. Flush eyes vigorously with water. Quickly don, clear and seal the mask.

3. Decontaminate your hands, face, and the inside of your mask.

   a. Remove one RSDL packet from your carrying pouch.
Safety: Avoid contact with eyes and wounds. If contact with eyes or wounds occurs, rinse with water as soon as possible.

   b. Tear it open quickly at any notch.

   c. Remove the applicator pad from the packet, and save the packet as the remaining lotion can be added to the applicator pad, if required.

   d. Thoroughly scrub the exposed skin of your hand, palm, and fingers with the applicator pad.
Note: The applicator pad can be used from either side and may gripped in any manner allowing the applicator pad to be applied to the skin.

   e. Switch the applicator pad to the other hand, and repeat the procedure.
Note: 1. Do not discard the applicator pad at this time.

2. If you were masked with your hood secured when you became contaminated, stop. Put on your protective gloves, and proceed to step 4.

3. If you were not masked with the hood secured when you became contaminated, continue decontaminating the exposed skin.

DANGER

Death or injury may result if you breathe toxic agents while doing the following steps. If you need to breathe before you finish, reseal your mask, clear it, check it, get your breath, and then resume the decontaminating procedure.

f. Stop breathing, close eyes, grasp mask beneath chin and pull mask away from chin enough to allow one hand between the mask and your face. Hold the mask in this position during steps (3g) through (3m).

g. Thoroughly scrub the exposed skin of your face with lotion from the applicator pad.

h. Thoroughly scrub across your forehead.

i. Beginning at one side, scrub up and down across your cheeks, nose, chin, and closed mouth. Avoid ingesting.

j. Scrub under the chin from the ear along the jawbone to the other ear to coat your skin with lotion.

k. Turn your hand over and scrub the inside surfaces of the mask that may touch your skin. Be sure to include the drinking tube.

CAUTION

Do not apply lotion to the lens of the protective mask. The RSDL may cause loss of transparency.

l. Keep the applicator.

m. Seal your mask immediately, clear it, and check it.

n. Use the applicator and any remaining lotion in the packet. Without breaking the mask seal, scrub the applicator pad across the forehead, exposed scalp, the skin of the neck, ears, and throat.

o. Secure the hood.

p. Thoroughly scrub your hands with lotion again as in steps (3d) through (3e).

q. Assume MOPP Level 4 by putting on protective gloves.

> **WARNING**
>
> Do not discard the RSDL packaging or applicator pads into containers that contain HTH or STB. Heat and/or fire may result.

4. Use any remaining lotion to spot decontaminate weapons, personal equipment, and canteen cap that may have become contaminated.

5. Allow RSDL to remain on skin for at least 2 minutes to destroy the chemical agent.

6. Discard the used packet(s) and applicator pad(s) by leaving them in place. *Note:* Do not put used packets in your pockets. Discard the carrying pouch after using the packets.

7. Remove the decontaminating lotion with soap and water when operational conditions permit, such as an "All Clear" directive or after detailed troop decontamination.

Note: Upon completion of training and evaluation, ensure that Soldiers have adequate mask cleaning supplies and water to clean training RSDL off of their protective mask.

8. Decontaminate your gloves, the exposed areas of your mask and hood, your weapon, and your helmet using the first mitt of the M295 Decontaminating Kit.

    a. Remove one decontamination packet from your pouch.

    b. Tear the packet open at any notch.

    c. Remove the decontamination mitt.

    d. Discard the empty packet.

    e. Unfold the decontamination mitt

    f. Grasp the green (nonpad) side of the decontamination mitt with your nondominant hand. Pat the other gloved hand with the decontamination mitt to start the flow of decontamination powder onto your glove. Rub your glove with the decontamination mitt until it is completely covered with decontaminating powder.

    g. Insert the decontaminated, gloved hand inside the decontamination mitt. Ensure that the pad side is in the palm of your hand and your thumb sticks through the appropriate thumbhole. Securely tighten the wristband on the gloved hand.

    h. Decontaminate individual equipment by rubbing with the pad side of the decontamination mitt until the equipment is thoroughly covered with decontamination powder. Pay special attention to areas that are hard to reach (such as cracks, crevices, and absorbent materials).

        (1) Decontaminate your other glove.

        (2) Decontaminate exposed areas of your mask and hood.

        (3) Decontaminate your weapon.

        (4) Decontaminate your helmet by patting it with the decontamination mitt

    i. Discard the decontamination mitt.

9. Decontaminate your LBE, IBA or IOTV and accessories, mask carrier, overboots, and gloves again using the second mitt in the M295 Decontamination Kit.

    a. Get another packet, and repeat steps (9a) through (9g). Then, perform the following:

        (1) Decontaminate load-carrying equipment (LCE), IBA, IOTV and accessories (such as canteen, ammo pouch, and first aid pouch).

        (2) Decontaminate your mask-carrying case.

        (3) Decontaminate your protective boots.

        (4) Repeat the decontamination process on your protective gloves.

    b. Discard the decontamination mitt.

    c. Get another packet and repeat steps (9a) through (9g). If liquid contamination is still suspected or detected. Rub or blot areas where contamination is still suspected or detected.

WARNING

The M295 kit only removes the liquid hazard. Decontaminated items may still present a vapor hazard. Do not unmask until it has been determined safe to do so.

---

10. Remove the decontaminating powder when operational conditions permit.

11. Notify your supervisor on the location of the used decontaminating materials, and await guidance on disposal procedures.

(Asterisks indicates a leader performance step.)

Evaluation Guidance: bsp; Score the Soldier GO if all performance measures are passed (P). Score the Soldier NO GO if any performance measure is failed (F). If the Soldier scores NO GO, show the Soldier what was done wrong and how to do it correctly.

Evaluation Preparation: bsp; Setup: Gather all items in the conditions statement so that they can be provided to the Soldier. A good time to evaluate this task is while in a field environment. Brief Soldier: Tell the Soldier that their exposed skin has been contaminated and they must decontaminate. Then tell them to decontaminate their personal equipment.

Evaluation Preparation:

Setup: Provide the Soldier with the items listed in the task conditions statement. A good time to evaluate this task is while in a field environment. Gather materials for the disposal of hazardous waste according to federal, state, and local rules and regulations.

Brief Soldier: Tell the Soldier what body parts and equipment are contaminated.

	Performance Measures	GO	NO GO
1	Donned protective mask and hood. Did not pull the drawstrings. Did not fasten the shoulder straps if hood had them.	⸻	⸻
2	Sought overhead cover or used a poncho for protection against further contamination.	⸻	⸻
3	Decontaminated hands, face, and the inside of mask.	⸻	⸻

a. Removed one RSDL packet from carrying pouch.

b. Tore open quickly at notch.

c. Removed the applicator pad from the packet, and saved the packet.

d. Thoroughly scrubbed the exposed skin hand, palm, and fingers with the applicator pad.

e. Switched the applicator pad to the other hand, and repeated the procedure.

f. Stopped breathing, closed eyes, grasped mask beneath chin and pulled mask away from chinenough to allow one hand between the mask and face. Held the mask in this position during steps (3g)through (3m).

g. Thoroughly scrubbed the exposed skin of the face with lotion from the applicator pad.

h. Thoroughly scrubbed across the forehead.

i. Beginning at one side, scrubbed up and down across cheeks, nose, chin, and closed mouth.

j. Scrubbed under the chin from the ear along the jawbone to the other ear coating skin with lotion.

k. Turned hand over and scrubbed the inside surfaces of the mask that may touch skin. Was sure tol. Kept the applicator.

l. Kept the applicator.

m. Sealed the mask immediately, cleared it, and checked it.

n. Used the applicator and any remaining lotion in the packet. Without breaking the mask seal, scrubbed the applicator pad across the forehead, exposed scalp, skin of the neck, ears, and throat

    o. Secured the hood.

    p. Thoroughly scrubbed hands with lotion again as in steps (3d) through (3e).

    q. Assumed MOPP Level 4 by putting on protective gloves. include the drinking tube.

4   Used any remaining lotion to spot decontaminate weapons, personal equipment, and canteen cap thatmay have become contaminated. _____ _____

5   Allowed RSDL to remain on skin for at least 2 minutes to destroy the chemical agent.

6   Discarded the used packet(s) and applicator pad(s) by leaving them in place. _____ _____

7   Removed the decontaminating lotion with soap and water when operational conditions permitted, suchas an "All Clear" directive or after detailed troop decontamination. _____ _____

8   Decontaminated gloves, the exposed areas of mask, hood, weapon, and helmet using the first mitt ofthe M295 Decontaminating Kit. _____ _____

    a. Removed one decontamination packet from pouch.

    b. Tore the packet open at the notch.c. Removed the decontamination mitt.d. Discarded the empty packet.

    e. Unfolded the decontamination mitt.

    f. Grasped the green (nonpad) side of the decontamination mitt with nondominant hand. Patted the otther gloved hand with the decontamination mitt to start the flow of decontamination powder onto glove. Rubbed glove with the decontamination mitt until it was completely covered with decontaminating powder.

    g. Inserted the decontaminated, gloved hand inside the decontamination mitt. Ensured that the padside is in the palm of hand and thumb stuck through the appropriate thumbhole.

Securely tightened thewristband on the gloved hand.

h. Decontaminated individual equipment by rubbing with the pad side of the decontamination mitt until the equipment was thoroughly covered with decontamination powder. Paid special attention to areas that are hard to reach (such as cracks, crevices, and absorbent materials).

h. Decontaminated individual equipment by rubbing with the pad side of the decontamination mitt until the equipment was thoroughly covered with decontamination powder. Paid special attention to areas that are hard to reach (such as cracks, crevices, and absorbent materials).

(1) Decontaminated the other glove.

(2) Decontaminated exposed areas of mask and hood.

(3) Decontaminated weapon.

(4) Decontaminated helmet by patting it with the decontamination mitt.

i. Discarded the decontamination mitt. Decontaminated LBE, IBA or IOTV and accessories, mask carrier, overboots, and gloves again using the second mitt of the M295 Decontaminating Kit.

9   a. Got another packet, and repeated steps (9a) through (9g). Then, performed the following

(1) Decontaminated load-carrying equipment (LCE), IBA, IOTV and accessories (such as canteen, ammo pouch, and first aid pouch).

(2) Decontaminated mask-carrying case.

(3) Decontaminated protective boots.

(4) Repeated the decontamination process on protective gloves.

b. Discarded the decontamination mitt.

c. Got another packet and repeated steps (9a) through (9g). If liquid contamination was still suspected or detected. Rubbed or blotted

Performance Measures	GO	NO GO
areas where contamination was still suspected or detected.		
10 Removed the decontaminating powder when operational conditions permitted.	_____	_____
11 Notified the supervisor on the location of the used decontaminating materials, and awaited guidanceon disposal procedures.	_____	_____

Evaluation Guidance: Score the Soldier GO if all performance measures are passed (P). Score the Soldier NO-GO if any performance measure is failed (F). If the Soldier fails any performance measure, show him how to do it correctly.

References:
Required TM 3-11.32, TM 10-8415-209-10, TM 10-8415-220-10, TM 3-4230-229-10, TM 3-4230-235-10

Related: Biological, and Chemical (NBC) Protection.

031-COM-1001

React to Nuclear Hazard/Attack

Foreign Disclosure: FD1 - This training product has been reviewed by the training developers in coordination with USACBRNS foreign disclosure (FD) representative and MCCoE foreign disclosure officer. This training product can be used to instruct international military students from all approved countries without restrictions.

Conditions: You are in an area where a nuclear weapons have been deployed. You are given a protective mask, a brush or a broom, and shielding material. Some iterations of this task should be performed in MOPP 4.

CAUTION

DO NOT USE MASKS WITH DAMAGED FILTERS BECAUSE CERTAIN MODELS CONTAIN HAZARDOUS MATERIALS. DO NOT CHANGE THE FILTER IN A CONTAMINATED ENVIRONMENT.

Standard: React to a nuclear attack by performing the steps in sequence without becoming a casualty.

Special Condition: None

Safety Level: Low

MOPP: Sometimes

Task Statements

Cue: 1. You observe a bright flash, 2. There is an enormous explosion, high winds, and a mushroom-shaped cloud forms, clearly indicating a nuclear attack, 3. Third-party information, including current intelligence is received (for example, terrorist warning or report of an incident at a nuclear facility by the Host Nation), or 4. You receive instructions to respond to a nuclear attack.

Performance Steps

1. Drop down immediately.
   a. If in Open Area, drop facedown immediately with feet facing the blast. *Note:* This will lessen the possibility of heat/blast injuries to the head, face, and neck.
      (1) If time, crawl to the closest available protection (i.e. A log, a large rock, or any depression in the earth's surface provides some protection).
      (2) If time, don IPE (individual protective equipment) according to the unit SOP.
   b. If in a Shelter, lay face down on the floor near a wall, if time, don IPE (individual protective equipment) according to the unit SOP.
   c. If in a Foxhole, the best position is on the back with knees drawn up to the chest, hands holding back of knees, if time, don IPE (individual protective equipment) according to the unit SOP. *Note:* This position may seem vulnerable, but the arms and legs are more radiation-resistant and will protect the head and trunk.

2. Close eyes and open mouth.
   Note: This will equalize the blast pressure and help prevent organ damage.

3. If not in a foxhole position, protect exposed skin from heat by putting hands and arms under or near the body. In any position keep the helmet on. *Note:* Doesn't apply to foxhole position.

4. If in Open Area, use any available material to provide overhead cover after the blast wave passes to avoid direct contact with radioactive fallout (rain gear, poncho, tarps, or plastic).

5. Remain in position until the blast wave passes and debris stops falling.

6. Decontaminate Yourself.
   a. Don protective mask or dust mask, if not already on.
   Note: This ensures that personnel protect themselves from ingesting or inhaling the radioactive particles.

b. Brush or shake debris off of clothing.

c. Lift off dry contamination with sticky tape if available.

d. Wash exposed skin with soap (detergent) and tepid water.

Note: This would be all skin that was exposed during the attack.

7. Check for casualties.

8. Seek shelter, if not already in a shelter/foxhole with overhead cover.

(Asterisks indicate a leader performance step.)

Evaluation Guidance: Read the action, condition, and standard to the Soldier. Provide the Soldier with all items given in the Conditions
Statement. Score the Soldier GO if all performance measures are passed (P) in sequence. Score the Soldier NO GO if any performance measure is failed (F) or out of sequence. If the Soldier scores NO GO, show the Soldier what was done wrong and how to do it correctly.

Evaluation Preparation: Gather all items in the conditions statement so that they can be provided to the Soldier. Prepare scenarios and questions to ask the Soldier in order to illicit a response that can be evaluated against the performance measures.

Performance Measures	GO	NO GO
1 Dropped down immediately.	_____	_____

Note : Donning protective mask first prevents Alpha particles from entering the nose, mouth, throat, and lungs.

a. If in Open Area, dropped facedown immediately with feet facing the blast.

(1) If time, crawled to the closest available protection (i.e. A log, a large rock, or any depression in the earth's surface provides some protection).

(2) If time, donned IPE (individual protective equipment) according to the unit SOP.

b. If in a Shelter, laid face down on the floor near a wall, if time, donned IPE (individual protective equipment) according to the unit SOP.

Performance Measures	GO	NO GO

    c. If in a Foxhole, laid on back with knees drawn up to the chest, hands holding back of knees, if time,

    donned IPE (individual protective equipment) according to the unit SOP.

2   Closed eyes and opened mouth.   _____ _____

    Note : This will equalize the blast pressure and help prevent organ damage.

3   In any position kept the helmet on.   _____ _____

4   4. If in Open Area, used any available material to   _____ _____
provide overhead cover after the blast wave passes to avoid direct contact with radioactive fallout (rain gear, poncho, tarps, or plastic).

5.   Remained in position until the blast wave passed and debris stopped falling.

    _____ _____

6.   Decontaminated your self.
    a. Donned protective mask or dust mask, if not already on.
    b. Brushed or shook debris off of clothing.
    c. Lifted off dry contamination with sticky tape if available.
    d. Washed exposed skin with hot soapy water. This would be all skin that was exposed during the attack.   _____ _____

7.   Checked for casualties.   _____ _____

8.   Sought shelter, if not already in a shelter/foxhole with overhead cover.   _____ _____

References:
Required: ATP 3-11.32, ATP 4-25.13, ECBC-SP-036, TM 10-8415-220-10, TM 3-4240-542-13&P
Related:

Environment: Environmental protection is not just the law but the right thing to do. It is a continual process and starts with deliberate planning. Always be alert to ways to protect our environment during training and missions. In doing so, you will contribute to the sustainment of our training resources while protecting people and the environment from harmful effects. Refer to the current Environmental Considerations manual and the current GTA Environmental-related Risk Assessment card.

Safety: In a training environment, leaders must perform a risk assessment in accordance with current Risk Management Doctrine. Leaders will complete the current Deliberate Risk Assessment Worksheet in accordance with the TRADOC Safety Officer during the planning and completion of each task and sub-task by assessing mission, enemy, terrain and weather, troops and support available-time available and civil considerations, (METT-TC). *Note:* During MOPP training, leaders must ensure personnel are monitored for potential heat injury. Local policies and procedures must be followed during times of increased heat category in order to avoid heat related injury. Consider the MOPP work/rest cycles and water replacement guidelines IAW current CBRN doctrine.

Supporting tasks:
071-121-4066 Prepare an Armor/cavalry vehicle for Nuclear Attack

---

031-COM-1003

Mark CBRN-Contaminated Areas

Foreign Disclosure: FD1 - This training product has been reviewed by the training developers in coordination with the USACBRNS Foreign Disclosure (FD) representative and MSCoE foreign disclosure officer. This training product can be used to instruct international military students from all approved countries without restrictions.

Conditions: You in an environment where CBRN weapons have been deployed. The contamination has been located and identified in an area. You are given a M328 Chemical, Biological, Radiological, and Nuclear (CBRN) marking kit, and appropriate Individual Protective Equipment (IPE). This task.

Standards: Mark the CBRN-contaminated area with the appropriate sign according to type of contamination and 100% of the required informationwritten on the sign. Emplace a minimum of three markers at line of sight distances depending on terrain. This task will be performed in mission-orientedprotective posture (MOPP) level 4 or Level A suit depending on CBRN material used.

Special Condition: None

Special Standards: None

Special Equipment:

Task Statements

Cue: 1. CBRN material is detected, 2. Given a requirement to mark an area known to be CBRN contaminated.

Remarks: MOPP 4 or appropriate level of protective equipment (Individual Protective Equipment (IPE) or Personal Protective Equipment (PPE)), based on the type of toxic material in the operational environment. If the type of toxic material in the operational environment is unknown, then the appropriate level of protective equipment should always be in Level A PPE.

Performance Steps

    1. Employ CBRN Markers (based on contamination type).

    a. Employ the ATOM marker for Radiological or Nuclear contamination.

        (1) Place markers at the location where a dose rate of 1 centigray per hour (cGyph) or more is measured.

        (2) Place markers so that the word "ATOM" faces away from the contamination at waist height right-angled apex downward.

        (3) Print the following information clearly on the front of the markers: *Note:* case of limited space on the front surface of the sign, as a minimum, the name/symbol of the agent (if known) and/or the dose rate/concentration (if known) is to be written on the front surface. Any other details may be written on the back surface.

            (a) Date-time group (DTG) (Local/Zulu (L/Z)) of reading. If the DTG is not known, print "unknown".

            (b) Dose rate.

            (c) DTG (L/Z) of detonation/release, if known. If the DTG is not known, print "unknown".

    b. Employ the ATOM marker for Toxic Industrial Radiological (TIR). *Note:* In case of limited space on the front surface of the sign, as a minimum, the name/symbol of the agent (if known) and/or the dose rate/concentration (if known) is to be written on the front surface. Any other details may be written on the back surface.

        (1) Place markers at the location where a dose rate of 2 micrograys per hour (μGyph) or more is measured.

        (2) Place markers so that the word "ATOM" faces away from the contamination at waist height right-angled apex downward.

        (3) Print the following information clearly on the front of the markers: *Note:* case of limited space on the front surface of the sign, as a minimum, the name/symbol of the agent (if known) and/or the doserate/concentration (if

known) is to be written on the front surface. Any other details may be written on the back surface.

        (a) DTG (L/Z) of reading. If the DTG is not known, print "unknown".

        (b) Dose rate.

        (c) DTG (L/Z) of detonation/release, if known. If the DTG is not known, print "unknown".

    c. Employ the BIO marker for Biological Agent's.

        (1) Place markers 200 meters before the location where contamination is detected.

        (2) Place markers so that the word "BIO" faces away from the contamination at waist height right-angled apex downward.

        (3) Print the following information clearly on the front of the markers:

            (a) Name of agent/symbol, if known. If unknown, print "unknown".

            (b) Concentration levels, if known. If unknown, print "unknown".

            (c) DTG (L/Z) of detection. If the DTG is not known, print "unknown".

            (d) DTG (L/Z) of detonation/release. If the DTG is not known, print "unknown".

    d. Employ the GAS marker for Persistent Chemical Agent's.

        (1) Place markers 200 meters before the location where contamination is detected.

        (2) Place markers so that the word "GAS" faces away from the contamination at waist height right-angled apex downward.

        (3) Print the following information clearly on the front of the markers:

            (a) Name of agent/symbol, if known. If unknown, print "unknown".

            (b) Concentration levels, if known. If unknown, print "unknown".

            (c) DTG (L/Z) of detection. If the DTG is not known, print "unknown".

            (d) DTG (L/Z) of detonation/release. If the DTG is not known, print "unknown".

    e. Employ the TOXIC marker for Toxic Industrial Chemical (TIC) or Toxic Industrial Biological (TIB).

        (1) Place markers 200 meters before the location where contamination is detected.

        (2) Place markers so that the word "TOXIC" faces away from the contamination at waist height right-angled apex downward.

        (3) Print the following information clearly on the front of the markers:

            (a) Name of agent/symbol, if known. If unknown, print "unknown".

            (b) Concentration levels, if known. If unknown, print "unknown".

            (c) DTG (L/Z) of detection. If the DTG is not known, print "unknown".

            (d) DTG (L/Z) of detonation/release. If the DTG is not known, print "unknown".

  2. Emplace two additional markers, at a minimum, using procedures from step 1

a. Place markers 10 to 100 meters apart, depending on terrain, ensuring all markers are line-of-sight visible.

b. When marking a contaminated area in open terrain (that is, desert, plains, rolling hills), raise the markers to a desired height that permits approaching military forces to view the markers at distances up to 200 meters.. (Asterisks indicates a leader performance step.)

Evaluation Guidance: Read the action, condition, and standard to the Soldier. Provide the Soldier with all items given in the Conditions Statement. Score the Soldier GO if all performance measures are passed (P) in sequence. Score the Soldier NO GO if any performance measure is failed (F) or out of sequence. If the Soldier scores NO GO, show the Soldier what was done wrong and how to do it correctly.

Evaluation Preparation: Gather all items in the conditions statement so that they can be provided to the Soldier. Prepare scenarios and questions to ask the Soldier in order to illicit a response that can be evaluated against the performance measures.

Evaluation Preparation:

Setup: Provide the Soldier with the items listed in the task condition statement. Use simulants to produce a contaminated environment for toxic and chemical or biological agents. For radiological contamination, tell the Soldier the type and amount of radiation present.

Brief Soldier: Tell the Soldier that the test will consist of ensuring that NBC markers are properly emplaced and that all required information is placed on the markers.

Performance Measures	GO	NO GO
1 Employed CBRN Markers (based on contamination type).	_____	_____
a. Employed the ATOM marker for Radiological or Nuclear contamination.		
(1) Placed markers at the location where a dose rate of 1 centigray per hour (cGyph) or more is measured.		

(2) Placed markers so that the word "ATOM" faces away from the contamination at waist height right-angled apex downward.

(3) Printed the following information clearly on the front of the markers: *Note:* In case of limited space on the front surface of the sign, as a minimum, the name/symbol of the agent (if known) and/or the dose rate/concentration (if known) is to be written on the front surface. Any other details may be written on the back surface.

(a) Date-time group (DTG) (Local/Zulu (L/Z)) of reading. If the DTG is not known, print "unknown".

(b) Dose rate.

(c) DTG (L/Z) of detonation/release, if known. If the DTG is not known, print "unknown".

b. Employed the ATOM marker for Toxic Industrial Radiological (TIR).

(1) Placed markers at the location where a dose rate of 2 micrograys per hour (μGyph) or more is measured.

(2) Placed markers so that the word "ATOM" faces away from the contamination at waist height right-angled apex downward.

(3) Printed the following information clearly on the front of the markers: *Note:* In case of limited space on the front surface of the sign, as a minimum, the name/symbol of the agent (if known) and/or the dose rate/concentration (if known) is to be written on the front surface. Any other details may be written on the back surface.

(a) DTG (L/Z) of reading. If the DTG is not known, print "unknown".

(b) Dose rate.

(c) DTG (L/Z) of detonation/release, if known. If the DTG is not known, print "unknown".

c. Employed the BIO marker for Biological Agent's.

(1) Placed markers 200 meters before the location where contamination is detected.

(2) Placed markers so that the word "BIO" faces away from the contamination at waist height right angled apex downward.

(3) Printed the following information clearly on the front of the markers: *Note*: In case of limited space on the front surface of the sign, as a minimum, the name/symbol of the agent (if known) and/or the dose rate/concentration (if known) is to be written on the front surface. Any other details may be written on the back surface.

(a) Name of agent/symbol, if known. If unknown, print "unknown".

(b) Concentration levels, if known. If unknown, print "unknown".

(c) DTG (L/Z) of detection. If the DTG is not known, print "unknown".

(d) DTG (L/Z) of detonation/release. If the DTG is not known, print "unknown".

d. Employed the GAS marker for Persistent Chemical Agent's.

(1) Placed markers 200 meters before the location where contamination is detected.

(2) Placed markers so that the word "GAS" faces away from the contamination at waist height rightangled apex downward.

(3) Printed the following information clearly on the front of the markers: *Note*: In case of limited space on the front surface of the sign, as a minimum, the name/symbol of the agent (if known) and/or the dose rate/concentration (if known) is to be written on the front surface. Any other details may be written on the back surface.

(a) Name of agent/symbol, if known. If unknown, print "unknown".

(b) Concentration levels, if known. If unknown, print "unknown".

(c) DTG (L/Z) of detection. If the DTG is not known, print "unknown".

(d) DTG (L/Z) of detonation/release. If the DTG is not known, print "unknown".

e. Employed the TOXIC marker for Toxic Industrial Chemical (TIC) or Toxic Industrial Biological (TIB).

(1) Placed markers 200 meters before the location where contamination is detected.

(2) Placed markers so that the word "TOXIC" faces away from the contamination at waist height right-angled apex downward.

(3) Printed the following information clearly on the front of the markers: *Note*: In case of limited space on the front surface of the sign, as a minimum, the name/symbol of the agent (if known) and/or the dose rate/concentration (if known) is to be written on the

front surface. Any other details may be written on the back surface.

(a) Name of agent/symbol, if known. If unknown, print "unknown".

(b) Concentration levels, if known. If unknown, print "unknown".

(c) DTG (L/Z) of detection. If the DTG is not known, print "unknown".

(d) DTG (L/Z) of detonation/release. If the DTG is not known, print "unknown".

2 Emplaced two additional markers, at a minimum, using procedures from step 1.   _____   _____

a. Placed markers 10 to 100 meters apart, depending on terrain, ensuring all markers are line-of-sight visible.

b. When marking a contaminated area in open terrain (that is, desert, plains, rolling hills), raised the markers to a desired height that permits approaching military forces to view the markers at distances up to 200 meters.

Evaluation Guidance: Score the Soldier GO if all performance measures are passed (P). Score the Soldier NO-GO if any performance measure is failed (F). If the Soldier fails any performance measure, show him how to do it correctly.

References:
Required: ATP 3-11.32, ATP 3-11.37, TM 3-9905-002-10
Related:

081-COM-1001

Evaluate a Casualty (Tactical Combat Casualty Care)

Conditions: While in a tactical area of operations, you encounter a combat casualty. Your unit may be under fire.

Some iterations of this task should be performed in MOPP.

Standards: Evaluate the casualty following the correct sequence. Identify and treat all life-threatening conditions and other serious wounds.

Special Condition: None

Special Standards: None

Safety Level: Low

MOPP 4: Sometimes

Special Equipment:

Task Statements

Cue: None
Note: All required references and technical manuals will be provided by the local command

Performance Steps

WARNING
If a broken neck or back is suspected, do not move the casualty unless to save his/her life.

1. Perform care under fire.
    a. Return fire as directed or required before providing medical treatment.
    b. Determine if the casualty is alive or dead.
Note: In combat, the most likely threat to the casualty's life is from bleeding. Attempts to check for airway and breathing will expose the rescuer to enemy fire. Do not attempt to provide first aid if your own life is in imminent danger. In a combat situation, if you find a casualty with no signs of life--no pulse, no breathing--do NOT attempt to restore the airway. Do NOT continue first aid measures.

c. Provide care to the live casualty. Direct the casualty to return fire, move to cover, and administer self-aid (stop bleeding), if possible.

Note: Reducing or eliminating enemy fire may be more important to the casualty's survival than the treatment you can provide.

If the casualty is unable to move and you are unable to move the casualty to cover and the casualty is still under direct enemy fire, have the casualty "play dead."

Cue: Enemy fire has been suppressed

d. In a battle-buddy team, approach the casualty (use smoke or other concealment if available using the most direct route possible.

e. Administer life-saving hemorrhage control.

(1) Determine the relative threat of enemy fire versus the risk of the casualty bleeding to death.

(2) If the casualty has severe, life-threatening bleeding from an extremity or has an amputation of an extremity, administer life-saving hemorrhage control by applying a tourniquet from the casualty's IFAK before moving the casualty. (See task 081-COM-1032.)

Note: The only treatment that should be given at the point of injury is a tourniquet to control life-threatening extremity bleeding.

f. Move the casualty, his weapon, and mission-essential equipment when the tactical situation permits.

g. Recheck bleeding control measures (tourniquet) as soon as behind cover and not under enemy fire.

Cue: You are now behind cover and are not under hostile fire.

2. Perform tactical field care.

Note: When evaluating and/or treating a casualty, seek medical aid as soon as possible. Do NOT stop treatment. If the situation allows, send another person to find medical aid.

a. Form a general impression of the casualty as you approach (extent of injuries, chance of survival).

Note: If a casualty is being burned, take steps to remove the casualty from the source of the burns before continuing evaluation and treatment. (See task 081-COM-1007.)

b. Check for responsiveness.

(1) Ask in a loud, but calm, voice: "Are you okay?" Gently shake or tap the casualty on the shoulder.

(2) Determine the level of consciousness by using AVPU: A = Alert; V = responds to Voice; P = responds to Pain; U = Unresponsive.

Note: To check a casualty's response to pain, rub the breastbone briskly with a knuckle or squeeze the first or second toe over the toenail. If casualty is wearing IBA, pinch his nose or his earlobe for responsiveness.

(3) If the casualty is conscious, ask where his body feels different than usual, or where it hurts.

Note: If the casualty is conscious but is choking and cannot talk, stop the evaluation and begin treatment. (See task 081-COM-1003.)

c. Identify and control bleeding.

(1) Check for bleeding.

(a) Reassess any tourniquets placed during the care under fire phase to ensure they are still effective.

(b) Perform a blood sweep of the extremities, neck, axillary, inguinal and extremity areas. Exposure is only necessary if bleeding is detected.

(1 Place your hands behind the casualty's neck and pass them upward toward the top of the head. *Note:* whether there is blood or brain tissue on your hands from the casualty's wounds.

(2 Place your hands behind the casualty's shoulders and pass them downward behind the back, the thighs, and the legs. Note whether there is blood on your hands from the casualty's wounds.

Note: If life-threatening bleeding is present, stop the evaluation and control the bleeding. (See task 081- COM-1032).

(2) Once bleeding has been controlled, continue to step 2d.

d. Position the casualty and open the airway. (See task 081-COM-1023.)

e. Assess for breathing and chest injuries.

(1) Expose the chest and check for equal rise and fall and for any wounds.

(2) Look, listen, and feel for respiration. (See task 081-COM-1023.)

Note: If the casualty is breathing, insert a nasopharyngeal airway (see task 081-COM-1023.) and place the casualty in the recovery position.

Only in the case of non-traumatic injuries such as hypothermia, near drowning, or electrocution should CPR be considered when in a tactical environment prior to the CASEVAC phase.

(3) If in a non-tactical environment, begin rescue breathing as necessary to restore breathing and/or pulse (See tasks 081-COM-1023 and 081-COM-0046.).

(a) If the casualty has a penetrating chest wound and is breathing or attempting to breathe, stop the evaluation to apply an occlusive dressing (See task 081-COM-1026.).

(b) Position or transport with the affected side down, if possible.

(c) Check for an exit wound. If found, apply an occlusive dressing.

f. Dress all non-life threatening injuries and any bleeding that has not been addressed earlier with appropriate dressings. (See task 081-COM-1032.)

3. Determine the need to evacuate the casualty and supply information for lines 3-5 of the 9-Line MEDEVAC request to your tactical leader. (See task 081-COM-0101.)

4. Check the casualty for burns.

a. Look carefully for reddened, blistered, or charred skin. Also check for singed clothes.

b. If burns are found, stop the evaluation and begin treatment. (See task 081-COM-1007.)

5. Administer pain medications and antibiotics (the casualty's combat pill pack) if available.

Note: Each Soldier will be issued a combat pill pack before deploying on tactical missions.

6. Document the injuries and the treatment given on the casualty's own Tactical Combat Casualty Care Card (found in IFAK), if applicable.

Performance Steps

Note: The FMC is usually initiated by the combat medic. However, a certified combat lifesaver can initiate the FMC if a combat medic is not available or if the combat medic directs the combat lifesaver to initiate the card. A pad of FMCs is part of the combat lifesaver medical equipment set.

    7. Transport the casualty to the evacuation site. (See task 081-COM-1046.)

    8. Monitor the patient for shock and treat as appropriate. (See task 081-COM-1005.) Continually reassess casualty until a medical person arrives or the patient arrives at a military treatment facility (MTF).

Evaluation Preparation:

Setup: Prepare a "casualty" for the Soldier to evaluate in step 2 by simulating one or more wounds or conditions. Simulate the wounds using a war wounds moulage set, casualty simulation kit, or other available materials. You can coach a "conscious casualty" on how to respond to the Soldier's questions about location of pain or other symptoms of injury. However, you will have to cue the Soldier during evaluation of an "unconscious casualty" as to whether the casualty is breathing and describe the signs or conditions, as the Soldier is making the checks.

Performance Measures	GO	NO GO
1 Performed care under fire.	_____	_____
2 Performed tactical field care.	_____	_____
3 Determined need to evacuate and reported information to tactical leader.	_____	_____
4 Checked the casualty for burns.	_____	_____
5 Administered pain medication and antibiotics (if applicable).	_____	_____
6 Documented injuries found on the Tactical Combat Casualty Card.	_____	_____
7 Transported the casualty to evacuation site.	_____	_____

Performance Measures	GO	NO GO
8 Monitored the patient for signs and symptoms of shock and treated as appropriate.	_____	_____

Evaluation Guidance: Tell the Soldier to do, in order, all necessary steps of Tactical Combat Casualty Care and treat all wounds and/or conditions identified appropriately. Tell the Soldier that he/she will not perform first aid but will tell you what first aid action (give mouth-to-mouth resuscitation, bandage the wound, and so forth) he/she would take. After he/she has completed the checks ask him/her what else should be done. To test step 8, ask the Soldier what should be? While evacuating an unconscious casualty. Tell the Soldier to do, in order, all necessary steps of Tactical Combat Casualty Care and treat all wounds and/or conditions identified appropriately. Tell the Soldier that he/she will not perform first aid but will tell you what first aid action (give mouth-to-mouth resuscitation, bandage the wound, and so forth) he/she would take. After he/she has completed the checks ask him/her what else should be done. To test step 8, ask the Soldier what should he be doing while evacuating an unconscious casualty.

References:
Required:
Related:

081-COM-1005

Perform First Aid to Prevent or Control Shock

Conditions: You have a casualty that is displaying one or more symptoms of shock. You have a field jacket or a poncho. The casualty is breathing and there is no uncontrolled bleeding. Some iterations of this task should be performed in MOPP 4.

Standards: Apply measures to prevent or treat shock without causing further injury to the casualty.

Special Condition: None

Special Standards: None

Special Equipment: None

Safety Risk: Low

MOPP 4: Sometimes

Task Statements

Cue: None

Note:

Performance Steps

1. Check the casualty for signs and symptoms of shock.
    a. Sweaty but cool skin.
    b. Pale skin.
    c. Restlessness or nervousness.
    d. Thirst.
    e. Severe bleeding.
    f. Confusion.
    g. Rapid breathing.
    h. Blotchy blue skin.
    i. Nausea and/or vomiting.
2. Position the casualty.

   a. Move the casualty under a permanent or improvised shelter to shade him from direct sunlight.

   b. Lay the casualty on his back unless a sitting position will allow the casualty to breathe easier.

WARNING

If the casualty has an un-splinted fractured leg, an abdominal wound, or a head or spinal injury, do not elevate the casualty's legs.

---

   c. Elevate the casualty's feet higher than the heart using a stable object so the feet will not fall.

WARNING

Do not loosen clothing if in a chemical area.

---

3. Loosen clothing at the neck, waist, or anywhere it is binding.
4. Prevent the casualty from getting chilled or overheated. Using a blanket or clothing, cover the casualty to avoid loss of body heat by wrapping completely around the casualty.

Note: Ensure no part of the casualty is touching the ground, as this increases loss of body heat.

5. Calm and reassure the casualty.

a. Take charge and show self-confidence.

b. Assure the casualty that he/she is being taken care of.

6. Watch the casualty closely for life-threatening conditions and check for other injuries, if necessary. Seek medical aid.

7. Seek medical aid.

Evaluation Preparation:

Setup: For training and evaluation, use another Soldier to simulate a patient in shock.

Brief Soldier: Tell the Soldier to treat the casualty to prevent or control shock.

Performance Measures	GO	NO GO
1 Checked the casualty for signs and symptoms of shock.	_____	_____
2 Positioned casualty correctly.	_____	_____
3 Loosened clothing at the neck, waist, or anywhere it was binding.	_____	_____
4 Prevented the casualty from chilling or overheating.	_____	_____
5 Calmed and reassured the casualty.	_____	_____
6 Watched the casualty closely for life-threatening conditions and checked for other injuries, if necessary. Sought medical aid.	_____	_____
7 Sought Medical Aid.	_____	_____

Evaluation Guidance: Score each Soldier according to the performance measures. Unless otherwise stated in the task summary, the Soldier must pass all performance measures to be scored GO. If the Soldier fails any steps, show the Soldier what was done wrong and how to do the task correctly.

Environment: Environmental protection is not just the law but the right thing to do. It is a continual process and starts with deliberate planning. Always be alert to ways to protect our environment during training and missions. In doing so, you will contribute to the sustainment of our training resources while protecting people and the environment from harmful effects. Refer to ATP 3-34.5 Environmental Considerations and GTA 05-08-002 ENVIRONMENTAL-RELATED RISK ASSESSMENT. Environmental protection is not just the law but the right thing to do. It is a continual process and starts with deliberate planning. Always be alert to ways to protect our environment during training and missions. In doing so, you will contribute to the sustainment of our training resources while protecting people and the environment from harmful effects. Refer to ATP 3-34.5 Environmental Considerations and GTA 05-08-002 ENVIRONMENTAL-RELATED RISK ASSESSMENT.

Safety: In a training environment, leaders must perform a risk assessment in accordance with ATP 5-19, Risk Management. Leaders will complete a DD Form 2977 DELIBERATE RISK ASSESSMENT WORKSHEET during the planning and completion of each task and sub-task by assessing mission, enemy, terrain and weather, troops and support available-time available and civil considerations, (METT-TC). *Note:* During MOPP training, leaders must ensure personnel are monitored for potential heat injury. Local policies and procedures must be followed during times of increased heat category in order to avoid heat related injury. Consider the MOPP work/rest cycles and water replacement guidelines IAW TM 3-11.32 Multi-Service Reference for Chemical, Biological, Radiological, and Nuclear Warning and Reporting and Hazard Prediction Procedures.

References:
Required:
Related: TC 4-02.1

081-COM-1023

Perform First Aid to Open the Airway

Conditions: While performing tactical field care. You see an adult casualty with difficulty breathing. You are not in a chemical environment. You will need a Nasopharyngeal airway (NPA). This task should not be trained in MOPP 4.

Standards: Open the casualty's airway using the head-tilt-chin-lift or the jaw thrust maneuver without causing further injury while observing of the rise and fall of the casualty's chest.

Special Condition: None

Safety Risk: Low

Special Equipment:

MOPP 4: Never

Task Statements

DANGER
Do NOT use the head tilt/chin lift method if a spinal or neck injury is suspected.

Performance Steps

WARNING
The casualty should be carefully rolled as a whole, so the body does not twist.

CAUTION
All body fluids should be considered potentially infectious. Always observe body substance isolation (BSI) precautions.

Notes: The proponent for this task is 68W Health Care Specialist.

WARNING
The casualty should be carefully rolled as a whole, so the body does not twist.

1. Position the casualty onto his back, arms down, face up, on a flat hard surface.
   a. Kneel beside the casualty
   b. Raise the arm nearest you and straighten it above the casualties head.
   c. Position the legs so that they are together and straight
   d. Support the back of the head and neck with the hand nearest them.
   e. Grasp the casualty under the far arm with the free hand

f. Pull steadily and evenly toward yourself, keeping the head and neck in line with the torso.

g. Roll the casualty as a single unit.

h. Place the casualty's arms at their side.

Cue: Casualty is unconscious, does not appear to be breathing, and is lying on his or her back.

2. Open the airway.

Note: If foreign material or vomit is in the mouth, remove it as quickly as possible.

DANGER
Do not use this method if a spinal or neck injury is suspected.

a. Head-tilt/chin-lift method.

(1) Kneel beside the casualty's head and shoulders.

(2) Place the palm of one hand on the casualty's forehead and the index and middle fingers of the other hand on the bony part of the jaw below the chin.

(3) Tilt the casualty's head backward gently.

Note: Do NOT use the thumb to lift.

Note: Do NOT completely close the casualty's mouth.

CAUTION:

Do NOT press deeply into the soft tissue under the chin with the fingers.

(4) Release pressure on the chin to allow the mouth to open slightly once the head is tilted backward.

Cue: Use this method if a spinal or neck injury is suspected.

b. Jaw-Thrust Method

Note: If you are unable to maintain an airway after the second attempt, use the head-tilt/chin-lift method.

(1) Kneel above the casualty's head (looking toward the casualty's feet).

(2) Rest your elbows on the ground or floor.

(3) Place on hand on each side of the casualty's lower jaw at the angle of the jaw, below the ears.

CAUTION

Do not tilt or rotate the casualty's head.

(4) Use the index and middle fingers to push the angles of the casualty's lower jaw forward.

Note: If the casualty's lips are still closed after the jaw has been moved forward, use your thumbs to retract the lower lip and allow air to enter the casualty's mouth.

3. Check for breathing.

a. While maintaining the open airway position, place an ear over the casualty's mouth and nose, looking toward the chest and stomach.

b. Look for the chest to rise and fall.

c. Listen for air escaping during exhalation.

d. Feel for the flow of air on the side of your face.

e. Count the number of respirations for 15 seconds, multiply that number by 4 to get the rate of breaths per minute.

CAUTION

DO NOT use the NPA if there is clear fluid (cerebrospinal fluid-DSF) coming from the ears or nose. This may indicate a skull fracture.

f. If the casualty is unconscious, if respiratory rate is less than 2 in 15 seconds, and/or if the casualty is making snoring or gurgling sounds, insert an NPA.

(1) Keep the casualty in a face-up position.

(2) Lubricate the tube of the NPA with water.

(3) Push the tip of the casualty's nose upward gently.

(4) Position the tube of the NPA so that the bevel (pointed end) of the NPA faces toward the septum (the partition inside the nose that separates the nostrils).

CAUTION

Never force the NPA into the casualty's nostril. If resistance is met, pull the tube out and attempt to insert it in the other nostril. If neither nostril will accommodate the NPA, place the casualty in the recovery position.

(5) Insert the NPA into the nostril and advance it until the flange rests against the nostril.

4. Place casualty in the recovery position, if breathing normally (12-20 breaths per minute).

Note: Place the person's arm that is nearest you at a right angle to their body, so it is bent at the elbow with the hand pointing upwards. Gently pick up their other hand with your palm against theirs (palm to palm). Now place the back of their hand onto their opposite cheek (for example, against their left cheek if it is their right hand). Keep your hand there to guide and support their head as you roll them. Use your other arm to reach across to the person's knee that is furthest from you, and pull it up so that their leg is bent and their foot is flat on the floor. Gently pull their knee towards you so they roll over onto their side, facing you. Their body weight should help them to roll over quite easily. Move the bent leg that is nearest to you, in front of their body so that it is resting on the floor. This position will help to balance them. Gently raise their chin to tilt their head back slightly, as this will open up their airway and help them to breathe.

FIGURE 081-COM-1023-1

5. Seek medical assistance, if casualty is not breathing.

Evaluation Guidance: Score each Soldier according to the performance measures in the evaluation guide. Unless otherwise stated in the task summary, the Soldier must pass all performance measures to be scored GO. If the Soldier fails any step, show what was done wrong and how to do it correctly.

Evaluation Preparation: You must evaluate the students on their performance of this task in a field condition related to the actual task.

	Performance Measures	GO	NO GO
1	Position the casualty onto his/her back, if necessary, and placed him/her on a hard, flat surface.	_____	_____
2	Opened the airway.	_____	_____
3	Checked for breathing.	_____	_____

Performance Measures	GO	NO GO
4 Place casualty in the recovery position, if breathing normally (12-20 breaths per minute)	_____	_____
5 Sought medical aid, if casualty is not breathing	_____	_____

Environment: Environmental protection is not just the law but the right thing to do. It is a continual process and starts with deliberate planning. Always be alert to ways to protect our environment during training and missions. In doing so, you will contribute to the sustainment of our training resources while protecting people and the environment from harmful effects. Refer to the current Environmental Considerations manual and the current GTA Environmental-related Risk Assessment card.

Safety: In a training environment, leaders must perform a risk assessment in accordance with risk management doctrine.

References:
Required:
Related: TC 4-02.1

081-COM-1054

Apply an Emergency Bandage

Conditions: While in a Tactical Field Care phase, you have a combat casualty with a gunshot wound to an extremity, radial pulses present, and no altered mental status. Bleeding is not sever enough to warrant a Combat Application tourniquet (CAT) and requires an Emergency Bandage. This task should not be trained in MOPP 4.

Standards: Control bleeding by applying an Emergency Bandage in 3 minutes or less.

Special Condition: None

Safety Level: Low

MOPP: Never

CAUTION

All body fluids should be considered potentially infectious. Always observe body substance isolation (BSI) precautions by wearing gloves and eye protection as a minimal standard of protection.

Performance Steps

1. Tame body substance isolation
2. Expose the wound.
3. Pack the wound with gauze. Gauze should extend 1-2 inches above the skin.

Note: You must verbalize holding pressure for a minimum of 3 minutes. You can pack the wound with Combat Gauze or Kerlix.

4. Place white portion of the dressing down covering all of the wound.
5. Wrap the elastic portion of the bandage around the extremity.
6. Insert elastic wrap into the pressure bar.
7. Pull bandage in opposite direction apply pressure with the pressure bar over the wound.
8. Continue to wrap the wound tightly ensuring all edges of the wound pad are covered.
9. Secure the closure bar to the bandage.
10. Monitor for continued bleeding.
11. If bleeding is controlled then secure the bandage with tape.

If bleeding reoccurs then apply a CAT tourniquet until bleeding stops, the distal pulse is absent and secure with tape.

(Asterisks indicates a leader performance step.)

Evaluation Guidance: Must complete steps 3 to 10 in 4 minutes.

Score each Soldier according to the performance measures in the evaluation guide. Unless otherwise stated in the task summary, the Soldier must pass all performance measures to be scored GO. If the Soldier fails any step, show what was done wrong and how to do it correctly.

Evaluation Preparation: You must evaluate the students on their performance of this task in a field condition related to the actual task.

Performance Measures	GO	NO GO
1 Take body substance isolation.	_____	_____

Performance Measures	GO	NO GO
2 Expose the wound.	_____	_____
3 Pack the wound with gauze. Gauze should extend 1-2 inches above the skin.	_____	_____
4 Place white portion of the dressing down covering all of the wound.	_____	_____
5 Wrap the elastic portion of the bandage around the extremity.	_____	_____
6 Insert elastic wrap into the pressure bar.	_____	_____
7 Pull bandage in opposite direction apply pressure with the pressure bar over the wound.	_____	_____
8 Continue to wrap the wound tightly ensuring all edges of the wound pad are covered.	_____	_____
9 Continue to wrap the wound tightly ensuring all edges of the wound pad are covered.	_____	_____
10 Secure the closure bar to the bandage, Monitor for continued bleeding.	--------	-------
11 If bleeding is controlled then secure the bandage w	---------	-------
If bleeding reoccurs then apply a CAT tourniquet until bleeding stops, the distal pulse is absent and secure with tape		

References: DD Form 1380 Tactical Combat Casualty Care (TCCC) Card
Required:

Environment: Environmental protection is not just the law but the right thing to do. It is a continual process and starts with deliberate planning. Always be alert to ways to protect our environment during training and missions. In doing so, you will contribute to the sustainment of our training resources while protecting people and the environment from harmful effects. Refer to the current Environmental Considerations manual and the current GTA Environmental-related Risk Assessment card.

Safety: In a training environment, leaders must perform a risk assessment in accordance with current Risk Management Doctrine. Leaders will complete the current Deliberate Risk Assessment Worksheet in accordance with the TRADOC Safety Officer during the planning and completion of each task and sub-task by assessing mission, enemy, terrain and weather, troops and support available-time available and civil considerations, (METT-TC). *Note:* During MOPP training, leaders must ensure personnel are monitored for potential heat injury. Local policies and procedures must be followed during times of increased heat category in order to avoid heat related injury. Consider the MOPP work/rest cycles and water replacement guidelines IAW current CBRN doctrine.

081-COM-0099

Apply a Hemostatic Dressing

Conditions: While in the Tactical Field Care phase, you encounter a casualty who is bleeding externally. The wound is either not amenable to a tourniquet or a tourniquet is in place and alternate means of hemorrhage control are necessary. This task should not be trained in MOPP 4.

Standards: Apply a hemostatic dressing to control bleeding without causing further harm the casualty IAW Tactical Combat Casualty Care (TCCC).

Special Conditions: None

Safety Risk: Low

MOPP 4: Never

Task Statements

CAUTION

Hemostatic dressings should be applied with at least 3 minutes of direct pressure (optional for XStat). Each dressing works differently, so if one fails to control bleeding, it may be removed and a fresh dressing of the same type or a different type applied. (*Note:* XStat is not to be removed in the field, but additional XStat, other hemostatic adjuncts, or trauma dressings may be applied over it.)

Remarks: For compressible (external) hemorrhage not amenable to limb tourniquet use or as an adjunct to tourniquet removal, use Combat Gauze as the CoTCCC hemostatic dressing of choice. Alternative hemostatic adjuncts: Celox Gauze, ChitoGauze, or XStat (Best for deep, narrow-tract junctional wounds).

Notes: For non-APD references contact your training NCO and or check with the MOS library.

Performance Steps

1. Don Body Substance Isolation (BSI).
2. Remove all clothing or equipment to obtain access to the wound.
3. Identify the point of bleeding within the wound.
   a. Remove any pooled blood from the wound cavity with your hand or a wad of cotton gauze.
   b. Locate the bleeding vessel(s).
4. Pack Combat Gauze directly over the source of bleeding.
5. Pack the wound with the entire dressing. More than one Combat Gauze may be required.
6. Apply direct pressure for 3 minutes.
   a. Periodically check the dressing to ensure proper placement and bleeding control.
   b. If the bandage becomes completely soaked through and there is still active bleeding, pack a second Combat Gauze into the wound.
7. Bandage wound to secure the dressing in place.
   a. If the wound cavity is deep, apply cotton gauze (either wad or rolled) over the dressing.
   b. Secure dressing in place with either an emergency bandage or an elastic bandage (See task # 081-000-0110).
8. Secure the bandage in place with tape.
9. Record Treatment on a Tactical Combat Casualty Care (TCCC) Card (DD Form 1380). See task # 081-000-0013.
10. Request Medical Evacuation (see task # 081-000-0120).

(Asterisks indicates a leader performance step.)

Evaluation Guidance: Score each Soldier according to the performance measures in the evaluation guide. Unless otherwise stated in the task summary, the Soldier must pass all performance measures to be scored GO. If the Soldier fails any step, show what was done wrong and how to do it correctly.

Evaluation Preparation: You must evaluate the students on their performance of this task in a field condition related to the actual task.

Performance Measures	GO	NO GO
1 Took BSI.	___	___
2 Removed all clothing or equipment to obtain access to the wound.	___	___
3 3. Identified the point of bleeding within the wound.	___	___
a. Removed any pooled blood from the wound cavity with your hand or a wad of cotton gauze.		
b. Located the bleeding vessel(s).		
4 Packed Combat Gauze directly over the source of bleeding.	___	___
5 Packed the wound with the entire dressing.	___	___
6 6. Applied direct pressure for 3 minutes.	___	___
a. Periodically checked the dressing to ensure proper location and bleeding control.		
b. Packed a second Combat Gauze into the wound if the bandage became completely soaked through and there was still active bleeding.		

Performance Measures	GO	NO GO
7 Bandaged wound to secure the dressing in place.	_____	_____
a. Applied cotton gauze (either wad or rolled) over the dressing if the wound cavity was deep.		
b. Secured dressing in place with either an emergency bandage or an elastic bandage.		
8 8. Secured the bandage in place with tape.	_____	_____
9 Recorded Treatment on a Tactical Combat Casualty Care (TCCC) Card (DD Form 1380).	_____	_____
10 Request Medical Evacuation	_____	_____

References: DD Form 1380 Tactical Combat Casualty (TCCC) Card; ISBN 9781284041750; TCCC Required:

Environment: Environmental protection is not just the law but the right thing to do. It is a continual process and starts with deliberate planning. Always be alert to ways to protect our environment during training and missions. In doing so, you will contribute to the sustainment of our training resources while protecting people and the environment from harmful effects. Refer to the current Environmental Considerations manual and the current GTA Environmental-related Risk Assessment card.

Safety: In a training environment, leaders must perform a risk assessment in accordance with current Risk Management Doctrine. Leaders will complete the current Deliberate Risk Assessment Worksheet in accordance with the TRADOC Safety Officer during the planning and completion of each task and sub-task by assessing mission, enemy, terrain and weather, troops and support available-time available and civil considerations, (METT-TC). *Note:* During MOPP training, leaders must ensure personnel are monitored for potential heat injury. Local policies and procedures must be followed during times of increased heat category in order to avoid heat related injury. Consider the MOPP work/rest cycles and water replacement guidelines IAW current CBRN doctrine.

Prerequisite Individual Tasks : **None** Supporting Individual Tasks : **None** Supported Individual Tasks : **None** Supported Collective Tasks : **None**

081-COM-0069

Apply an Occlusive Dressing

Conditions: While in the Tactical Field Care phase, you encounter a casualty with an open chest wound. This task should not be trained in MOPP 4

Standards: Apply an Occlusive Dressing in 3 min or less.

Special Conditions: None

Safety Risk: Low

MOPP 4: Never

| **Task Statements** |

Remarks: None

Notes: For non-APD references contact your training NCO and or check with the MOS library.

Performance Steps

1. Take body substance isolation.
2. Expose and assess injury.

Note: Remove enough clothing to obtain access to the injury.

3. Upon full expiration, cover the wound with large, occlusive material dressing. (Cover the first wound encountered).

4. Log roll the casualty or have the conscious casualty sit up and examine the back for an exit wound.

5. If present, cover the exit wound on expiration with a large, occlusive dressing.

Note: Ensure material extends 2" beyond the edge of the wound. If improvised seal is used, tape four sides of occlusive dressing down.

6. Place casualty in sitting position or injured side down in the recovery position.

7. Verbalize continued assessment of casualty for signs of progressive respiratory distress.

(Asterisks indicates a leader performance step.)

Evaluation Guidance: Score each Soldier according to the performance measures in the evaluation guide. Unless otherwise stated in the task summary, the Soldier must pass all performance measures to be scored GO. If the Soldier fails any step, show what was done wrong and how to do it correctly.

Evaluation Preparation: You must evaluate the students on their performance of this task in a field condition related to the actual task.

Performance Measures	GO	NO GO
1 Took body substance isolation.	_____	_____
2 Exposed and assessed injury.	_____	_____
3 Upon full expiration, covered the wound with large, occlusive material dressing. (Cover the first wound encountered).	_____	_____
4 Log rolled the casualty or had the conscious casualty sit up and examined the back for an exit wound	_____	_____
5 If present, covered the exit wound on expiration with a large, occlusive dressing.	_____	_____
6 Placed casualty in sitting position or injured side down in the recovery position.	_____	_____
7 Verbalized continued assessment of casualty for signs of progressive respiratory distress.	_____	_____

References: ISBN 9781284041750
Required:
Environment: Environmental protection is not just the law but the right thing to do. It is a continual process and starts with deliberate planning. Always be alert to ways to protect our environment during training and missions. In doing so, you will contribute to the sustainment of our training resources while protecting people and the environment from harmful effects. Refer to the current Environmental Considerations manual and the current GTA Environmental-related Risk Assessment card.

Safety: In a training environment, leaders must perform a risk assessment in accordance with current Risk Management Doctrine. Leaders will complete the current Deliberate Risk Assessment Worksheet in accordance with the TRADOC Safety Officer during the planning and completion of each task and sub-task by assessing mission, enemy, terrain and weather, troops and support available-time available and civil considerations, (METT-TC). *Note:* During MOPP training, leaders must ensure personnel are monitored for potential heat injury. Local policies and procedures must be followed during times of increased heat category in order to avoid heat related injury. Consider the MOPP work/rest cycles and water replacement guidelines IAW current CBRN doctrine.

081-COM-0048

Apply a Combat Application Tourniquet (CAT)

Conditions: While in the Tactical Field Care phase, you encounter a casualty with life-threatening bleeding from an extremity. This task should not be trained in MOPP 4.

Standards: Apply a Combat Application Tourniquet in 60 seconds or less.

Special Conditions: None

Safety Risk: Low

MOPP 4: Never

Task Statements

CAUTION
All body fluids should be considered potentially infectious. Always observe body substance isolation (BSI) precautions by wearing gloves and eye protection as a minimal standard of protection.

Remarks: None

Notes: For non-APD references contact your training NCO and or check with the MOS library.

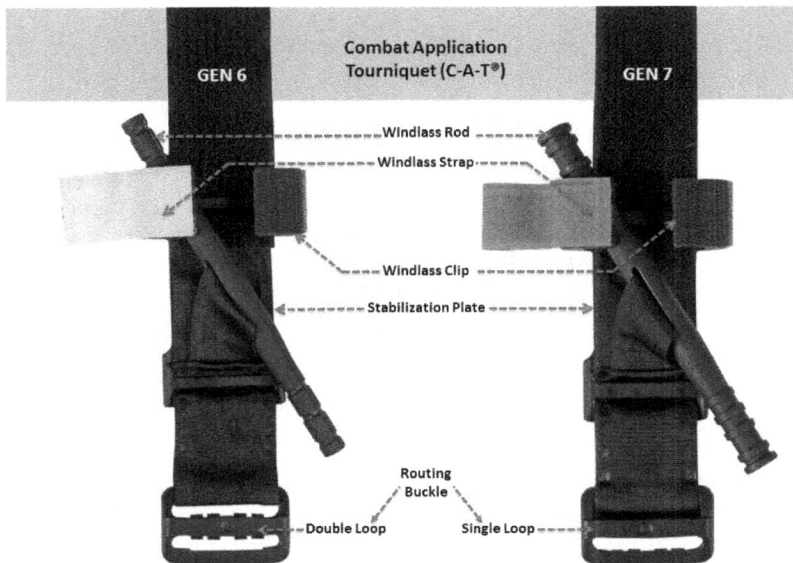

Figure 081-COM-0048-1

Picture showing the difference between the GEN 6 and GEN 7 CAT

Performance Steps

1. Take body substance isolation.
2. Expose, assess, and check for an exit wound.
3. Route the band around the limb above the wound on the injured extremity.
4. Pass the red tip through the inside slit in the buckle, position the CAT 2-3 inches above the wound, and directly on the skin.
5. Pull the band as tight as possible and secure the Velcro back on itself all the way around the limb, but not over the rod clips.

Note: Band should be tight enough that three finger tips cannot be slid between the band and the limb. If the tips of three fingers slide under the band, retighten and re-secure. This is the most important step.

6. Twist the windlass until the bleeding stops. (Should occur within 3 rotations of the windlass)

Cue: If the CAT was applied correctly the evaluator will state: "hemorrhage has been controlled". If the CAT was NOT applied correctly the evaluator will state: "hemorrhage HAS NOT been controlled".

7. Secure the windlass rod inside the windlass clip to lock it into place.
8. Check for distal pulse.

Note: If distal pulse is present; 1. Attempt additional tightening 2. If distal pulse is still present, apply a second CAT above and side by side with the first one.

9. If possible continue to route the self-adhering band between the windlass clips and over the windlass rod. Secure the rod and band with the windlass strap.

10. Place a "T" and the time of the application on the casualty.

11. Secure the CAT in place with tape.

12. Record the treatment on a DD Form 1380 Tactical Combat Casualty Care (TCCC) Card.

(Asterisks indicates a leader performance step.)

Evaluation Guidance: Score each Soldier according to the performance measures in the evaluation guide. Unless otherwise stated in the task summary, the Soldier must pass all performance measures to be scored GO. If the Soldier fails any step, show what was done wrong and how to do it correctly.

Evaluation Preparation: You must evaluate the students on their performance of this task in a field condition related to the actual task.

Performance Measures	GO	NO GO
1 Took body substance isolation.	_____	_____
2 Exposed, assessed, and checked for a wound.	_____	_____
3 Routed the band around the limb above the wound on the injured extremity.	_____	_____
4 Passed the red tip through the inside slit in the buckle, positioned the CAT 2-3 inches above the wound, and directly on the skin.	_____	_____
5 Pulled the band as tight as possible and secured the Velcro back on itself all the way around the limb, but not over the rod clips.	_____	_____
6 Twisted the windlass until the bleeding stops. (Should occur within 3 rotations of the windlass).	_____	_____

Performance Measures	GO	NO GO
7 Secured the windlass rod inside the windlass clip to lock it into place.	_____	_____
8 Checked for distal pulse.	_____	_____
9 If possible continued to route the self-adhering band between the windlass clips and over the windlass rod.	_____	_____
10 Secured the rod and band with the Placed a "T" and the time of the application on the casualty windlass strap.	--------	-------
11	--------	--------
12 Secured the CAT in place with tape.	--------	--------
Recorded the treatment on a DD Form 1380 Tactical Combat Casualty (TCCC) Card.		

References: DD FORM 1380 Tactical Combat Casualty Care (TCCC) Card; ISBN 9781284041750

Required:

Environment: Environmental protection is not just the law but the right thing to do. It is a continual process and starts with deliberate planning. Always be alert to ways to protect our environment during training and missions. In doing so, you will contribute to the sustainment of our training resources while protecting people and the environment from harmful effects. Refer to the current Environmental Considerations manual and the current GTA Environmental-related Risk Assessment card.

081-COM-0055

Apply a FOX Eye Shield

Conditions: While in the Tactical Field Care (TFC) phase, you encounter a casualty with an ocular injury. A visual acuity exam has been performed. This task should not be trained in MOPP 4.

Standards: Apply a Fox Eye Shield, without causing further injury IAW Tactical Combat Casualty Care (TCCC).

Special Conditions: None

Safety Risk: Low

MOPP 4: Never

Task Statements

Performance Steps
1. Position the casualty with head supported.
2. Ask the casualty to close both eyes.
3. Place the Fox eye shield over the injured eye.

Note: The Fox eye shield is designed to rest on the bony support of the face arching over the ocular structures.

4. Firmly secure the Fox eye shield with one or more strips of tape to the casualty's cheek and forehead.

Note: In the absence of a Fox eye shield other objects such as: Sam splint, Styrofoam or specimen cups can effectively perform the same function

---

CAUTION

To ensure that your casualty does not have an allergic reaction or go in to anaphylactic shock, check medication allergies by asking the casualty if they have any medication allergies or by checking for allergy tags.

---

5. Administer the 400 mg Moxifloxacin tablet(s) from the casualty's combat pill pack.

Note: Moxifloxacin is the antibiotic found in the casualty's combat pill pack, it should only be administered to a casualty that is conscious and able to swallow.

6. Document treatment and medication administered to the casualty on the DD Form 1380 Tactical Combat Casualty Care (TCCC) Card.

Note: The DD Form 1380 Tactical Combat Casualty Care (TCCC) Card, must be transported with the casualty to the next echelon to ensure continuity of care.

(Asterisks indicates a leader performance step.)

Evaluation Guidance: Score each Soldier according to the performance measures in the evaluation guide. Unless otherwise stated in the task summary, the Soldier must pass all performance measures to be scored GO. If the Soldier fails any step, show what was done wrong and how to do it correctly.

Evaluation Preparation: You must evaluate the students on their performance of this task in a field condition related to the actual task.

Performance Measures	GO	NO GO
1 Positioned the casualty with head supported.	_____	_____
2 Asked the casualty to close both eyes.	_____	_____
3 Placed the Fox eye shield over the injured eye.	_____	_____
4 Secured the Fox Eye shield with one or more strips of tape to the casualty's cheek and forehead.	_____	_____
5 Administered the 400 mg Moxifloxacin tablet(s) from the casualty's combat pill pack.	_____	_____
6 Documented treatment and medication administered to the casualty on the DD Form 1380 Tactical Combat Casualty Care (TCCC) Card.	_____	_____

TADSS: None

081-COM-0013

Initiate a DD Form 1380 Tactical Combat Casualty Care (TCCC) Card

Conditions: You are in a tactical combat environment. You have a casualty that has been medically treated and you must record the information on a Department of Defense (DD) Form 1380 Tactical Combat Casualty (TCCC) Card. You will need a DD Form 1380 and a pen. Some iterations of this task should be performed in MOPP 4.

Standards: Initiate the DD Form 1380 in accordance with Army Regulation (AR) 40-66, Medical Record Administration and Health Care Documentation.

Special Conditions: None

Safety Risk: Low

MOPP 4: S o m e t i m e s

Task Statements

CAUTION

All body fluids should be considered potentially infectious. Always observe body substance isolation (BSI) precautions by wearing gloves and eye protection as a minimal standard of protection.

Remarks: None

Notes: For non-APD references contact your training NCO and or check with the MOS library.

Performance Steps

1. Remove DD Form 1380 from the casualty's improved first aid kit.
2. Complete all entries as fully as possible

Note: As the DD Form 1380 is the first, and sometimes only, record of treatment of combat casualties, accuracy and thoroughness of information provided is of the utmost importance.

    a. Front of DD Form 1380.

      (1) Battle Roster # - Write first letter of casualty's first name, then first letter of casualty's last name, then write the last four numbers of casualty's Social Security number. For example, John Doe 123-12-1234 is Battle Roster # "JD1234".

      (2) Evacuation (EVAC) - Mark an "X" on the casualty's evacuation priority/precedence (Urgent; Priority; or Routine).

      (3) Name - Write casualty's name (Last, First).

      (4) Last 4 - Write last four numbers of casualty's Social Security number.

      (5) Gender - Mark an "X" on the casualty's gender Male (M) or Female (F).

      (4) Last 4 - Write last four numbers of casualty's Social Security number.

      (5) Gender - Mark an "X" on the casualty's gender Male (M) or Female (F).

      (6) Date - Write date of injury in DD-MMM-YY format. For example, "29-JUN-13".

      (7) Time - Write 24 hour time of injury, and indicate whether local (L) or zulu (Z) time. For example, "1300Z".

      (8) Service - Write casualty's branch of service (USA, USAF, USCG, USN, USMC). For U.S. civilians, write "US CIV". For non-U.S. personnel, write "NON US or a standard abbreviation for casualty's nationality.

      (9) Unit - Write casualty's unit name

(10) Allergies - Write casualty's known drug allergies. If no drug allergies, write no known drug allergies (NKDA).

(11) Mechanism of Injury - Mark an "X" on the mechanism or cause of injury (artillery, blunt, burn, fall, grenade, gunshot wound (GSW), improvised explosive device (IED), landmine, motor vehicle crash/collision (MVC), rocket-propelled grenade (RPG), other (specify)). Mark all that apply.

(12) Injury - Mark an "X" at the site of the injury(ies) on the body picture. For burn injuries, circle the burn percentage(s) on the figure. If multiple mechanisms of injury and multiple injuries, draw a line between the mechanism of injury and the anatomical site of the injury.

(13) TQ: R Arm (tourniquet, right arm) - If a tourniquet is applied to the right arm, write type of tourniquet used and the time of tourniquet application.

(14) TQ: L Arm (tourniquet, left arm) - If a tourniquet is applied to the left arm, write type of tourniquet used and the time of tourniquet application.

(15) TQ: R Leg (tourniquet, right leg) - If a tourniquet is applied to the right leg, write type of tourniquet used and the time of tourniquet application.

(16) TQ: L Leg (tourniquet, left leg) - If a tourniquet is applied to the left leg, write type of tourniquet used and the time of tourniquet application.

(17) Time - Write time of vital signs taken.

(18) Pulse (rate & location)- Write casualty's pulse rate.

(19) Blood Pressure - Write casualty's blood pressure

(20) Respiratory Rate - Write casualty's respiratory rate.

(21) Pulse Ox % O2 Sat - Write casualty's pulse Ox% and O2 saturation.

(22) AVPU - Write casualty's level of consciousness (AVPU: Alert, responds to Verbal stimulus, responds to Pain stimulus, Unresponsive).

(23) Pain Scale (0-10) - Write casualty's level of pain in numeric rating scale of 0 to 10, with 0 being no pain and 10 being the worst pain.

b. Back of Card.

(1) Battle Roster # - Write first letter of casualty's first name, then first letter of casualty's last name, and then write the last four numbers of casualty's Social Security number. For example, John Doe 123-12-1234 is Battle Roster # "JD1234".

(2) Evacuation (EVAC) - Mark an "X" on the casualty's evacuation priority/precedence (Urgent; Priority; or Routine).

(3) C - Mark an "X" for all Circulation hemorrhage control interventions. For tourniquets (TQ), mark category (Extremity, Junctional and/or Truncal) and write name of TQ(s) used. For dressings, mark category (Hemostatic, Pressure, and/or Other) and write type of dressing(s) used.

(4) A - Mark an "X" for all Airway interventions (Intact, nasopharyngeal airway (NPA), cricothyroidotomy (CRIC), endotracheal(ET)tube , supraglottic airway (SGA) and write type of device(s) used.

(5) B - Mark an "X" for all Breathing interventions oxygen (O2), needle decompression (Needle-D), Chest-Tube, (Chest-Seal) and write type of device(s) used.

(6) C: Fluid - Circulation resuscitation interventions. Write name, volume, route, and time of any fluids given.

(7) C: Blood Product - Circulation resuscitation interventions. Write name, volume, route, and time of any blood products given.

Note: When more space is needed for documentation, attach another DD Form 1380 to the original by safety pin or other means. The second form will be labeled DD Form 1380 #2 and will show the Soldier's name and unit.

(8) Meds: Analgesic Medications - Write name, dose, route, and time of any analgesics given.

(9) Meds: Antibiotic Medications - Write name, dose, route, and time of any antibiotics given.

(10) Meds: Other - Medications. Write name, dose, route, and time of any other administered medications.

(11) Other - Mark an "X" for other treatments administered (combat pill pack, eye shield (mark right (R) or left (L)), splint, hypothermia prevention) and type of device(s) used.

(12) Notes - Use this space to record any other pertinent information and/or clarifications.

(13) First Responder Name - Print the first responder's name (Last, First).

(14) First Responder Last 4 - Write last four numbers of first responder's Social Security number.

TACTICAL COMBAT CASUALTY CARE (TCCC) CARD

BATTLE ROSTER #: _____

EVAC: ☐ Urgent ☐ Priority ☐ Routine

NAME (Last, First): _____ **LAST 4:** _____

GENDER: ☐ M ☐ F **DATE** (DD-MMM-YY): _____ **TIME:** _____

SERVICE: _____ **UNIT:** _____ **ALLERGIES:** _____

Mechanism of Injury: (X all that apply)
☐ **Artillery** ☐ **Blunt** ☐ **Burn** ☐ **Fall** ☐ **Grenade** ☐ **GSW** ☐ **IED**
☐ **Landmine** ☐ **MVC** ☐ **RPG** ☐ **Other:** _____

Injury: (Mark injuries with an X)

TQ: R Arm	TQ: L Arm
TYPE:	TYPE:
TIME:	TIME:

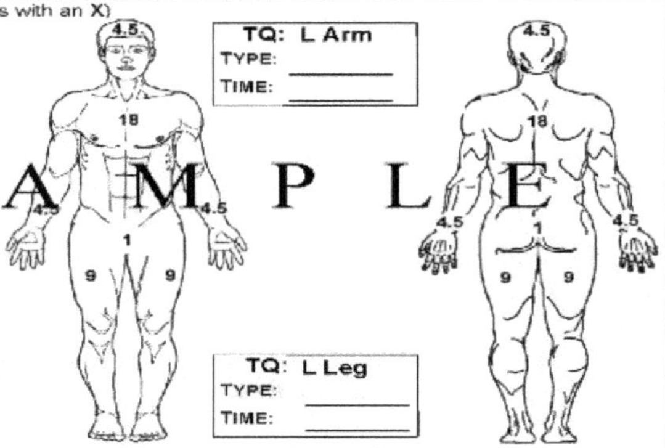

S A M P L E

TQ: R Leg	TQ: L Leg
TYPE:	TYPE:
TIME:	TIME:

Signs & Symptoms: (Fill in the blank)

	Time			
Pulse (Rate & Location)				
Blood Pressure	/	/	/	/
Respiratory Rate				
Pulse Ox % O2 Sat				
AVPU				
Pain Scale (0-10)				

DD Form 1380, JUN 2014   TCCC CARD

Figure 081-COM-0013-1 DD Form 1380 Page 1

Performance Steps

BATTLE ROSTER #: ______________________

EVAC: ☐ Urgent ☐ Priority ☐ Routine

Treatments: (X all that apply, and fill in the blank) *Type*

C: TQ- ☐ Extremity ☐ Junctional ☐ Truncal ____________

Dressing- ☐ Hemostatic ☐ Pressure ☐ Other ____________

A: ☐ Intact ☐ NPA ☐ CRIC ☐ ET-Tube ☐ SGA ____________

B: ☐ O2 ☐ Needle-D ☐ Chest-Tube ☐ Chest-Seal ____________

C:

	Name	*Volume*	*Route*	*Time*
Fluid				
Blood Product				

MEDS:

	Name	*Dose*	*Route*	*Time*
Analgesic (e.g., Ketamine, Fentanyl, Morphine)				
Antibiotic (e.g., Moxifloxacin, Ertapenem)				
Other (e.g., TXA)				

SAMPLE

OTHER: ☐ Combat-Pill-Pack ☐ Eye-Shield (☐ R ☐ L) ☐ Splint

☐ Hypothermia-Prevention Type: ____________

NOTES:

FIRST RESPONDER

NAME (Last, First): ______________________ LAST 4: __________

DD Form 1380, JUN 2014 (Back) TCCC CARD

Figure 081-COM-0013-2 DD Form 1380 Page 2

3. Attach the completed DD Form 1380 to the casualty; in a visible area where follow on medical staff will see it.

Performance Steps

Note: Do not attach the DD Form 1380 to the casualty's body armor as this equipment may will be separated from the casualty once they arrive at the medical treatment facility (MTF).

(Asterisks indicates a leader performance step.)

Evaluation Guidance: Score each Soldier according to the performance measures in the evaluation guide. Unless otherwise stated in the task summary, the Soldier must pass all performance measures to be scored GO. If the Soldier fails any step, show what was done wrong and how to do it correctly.

Evaluation Preparation: You must evaluate the students on their performance of this task in a field condition related to the actual task.

Performance Measures	GO	NO GO
1 Removed the DD Form 1380 from the casualty's improved first aid kit.	____	____
2 Completed all entries as fully as possible.	____	____
3 Attached the completed DD Form 1380 to the casualty; in a visible area where follow on medical staff will see it.	____	____

References: AR 40-400; AR 40-66; DD FORM 1380 Tactical Combat Casualty Care (TCCC) Card

Materiel Items (NSN):

Environment: Environmental protection is not just the law but the right thing to do. It is a continual process and starts with deliberate planning. Always be alert to ways to protect our environment during training and missions. In doing so, you will contribute to the sustainment of our training resources while protecting people and the environment from harmful effects. Refer to the current Environmental Considerations manual and the current GTA Environmental-related Risk Assessment card.

Safety: In a training environment, leaders must perform a risk assessment in accordance with current Risk Management Doctrine. Leaders will complete the current Deliberate Risk Assessment Worksheet in accordance with the TRADOC Safety Officer during the planning and completion of each task and sub-task by assessing mission, enemy, terrain and weather, troops and support available-time available and civil considerations, (METT-TC). *Note:* During MOPP training, leaders must ensure personnel are monitored for potential heat injury. Local policies and procedures must be followed during times of increased heat category in order to avoid heat related injury. Consider the MOPP work/rest cycles and water replacement guidelines IAW current CBRN doctrine.

081-COM-1046

Transport a Casualty

Conditions: You have a casualty who has received treatment and requires movement and/or evacuation from a vehicle and placement on a transportation platform. You may have assistance from other Soldiers. You will need materials to improvise a litter (poncho, shirts, or jackets, and poles or tree limbs), a SKED or Talon litter, and a vehicle or replicated platform to load patients onto. Some iterations of this task should be performed in MOPP.

Standards: Transport the casualty using an appropriate carry or litter without dropping or causing further injury to the casualty.

Special Condition: None

Safety Level: Low

MOPP 4: Sometimes

Task Statements

Cue: A casualty must be moved.

Note: N/A

Performance Steps

WARNING

If the casualty was involved in a vehicle crash you should always consider that he/she may have a spinal injury. Unless there is an immediate life-threatening situation (such as fire, explosion), do NOT move the casualty with a suspected back or neck injury. Seek medical personnel for guidance on how to transport the casualty.

1. Remove a casualty from a vehicle, if necessary.

a. Laterally.

(1) With the assistance of another Soldier grasp the casualty's arms and legs.

(2) While stabilizing the casualty's head and neck as much as possible, lift the casualty free of the vehicle and move him/her to a safe place on the ground.

Note: If medical personnel are available, they may stabilize the casualty's head, neck, and upper body with a special board or splint.

b. Upward.

Note: You may have to remove a casualty upward from a vehicle; for example, from the passenger compartment of a wheeled vehicle lying on its side or from the hatch of an armored vehicle sitting upright.

(1) You may place a pistol belt or similar material around the casualty's chest to help pull him/her from the vehicle.

(2) With the assistance of another Soldier inside the vehicle, draw the casualty upward using the pistol belt or similar material or by grasping his/her arms.

(3) While stabilizing the casualty's head and neck as much as possible, lift the casualty free of the vehicle and place him/her on the topmost side of the vehicle.

Note: If medical personnel are available, they may stabilize the casualty's head, neck, and upper body with a special board or splint.

(4) Depending on the situation, move the casualty from the topmost side of the vehicle to a safe place on the ground.

WARNING

Do NOT use manual carries to move a casualty with a neck or spine injury, unless a life-threatening hazard is in the immediate area. Seek medical personnel for guidance on how to move and transport the casualty.

---

2. Select an appropriate method to transport the casualty.

Note: The fireman's carry is the typical one-man carry practiced in training. However, in reality, with a fully equipped casualty, it is nearly impossible to lift a Soldier over your shoulder and move to cover quickly. It should be discouraged from being practiced and used.

a. Fireman's carry. Use for an unconscious or severely injured casualty.

CAUTION

Do NOT use the neck drag if the casualty has a broken arm or a suspected neck injury.

---

b. Neck drag. Use in combat, generally for short distances.

c. Cradle-drop drag. Use to move a casualty who cannot walk when being moved up or down stairs.

d. Use litters if materials are available, if the casualty must be moved a long distance, or if manual carries will cause further injury.

Cue: The appropriate type of carry has been selected.

3. Evacuate the casualty using a manual carry.

a. Fireman's carry.

(1) Kneel at the casualty's uninjured side.

(2) Place casualty's arms above his/her head.

(3) Cross the ankle on the injured side over the opposite ankle.

(4) Place one of your hands on the shoulder farther from you and your other hand on his/her hip or thigh.

(5) Roll the casualty toward you onto his/her abdomen.

(6) Straddle the casualty.

Note: This method is used if the rescuer believes that it is safer than the regular method due to the casualty's wounds. Care must be taken to keep the casualty's head from falling backward, resulting in a neck injury.

(7) Place your hands under the casualty's chest and lock them together.

(8) Lift the casualty to his/her knees as you move backward.

(9) Continue to move backward, thus straightening the casualty's legs and locking the knees.

(10) Walk forward, bringing the casualty to a standing position but tilted slightly backward to prevent the knees from buckling.

(11) Maintain constant support of the casualty with one arm. Free your other arm, quickly grasp his/her wrist, and raise the arm high.

(12) Instantly pass your head under the casualty's raised arm, releasing it as you pass under it.

(13) Move swiftly to face the casualty.

(14) Secure your arms around his/her waist.

(15) Immediately place your foot between his/her feet and spread them (approximately 6 to 8 inches apart).

(16) Again grasp the casualty's wrist and raise the arm high above your head.

(17) Bend down and pull the casualty's arm over and down your shoulder bringing his/her body across your shoulders. At the same time pass your arm between the legs.

(18) Grasp the casualty's wrist with one hand while placing your other hand on your knee for support.

(19) Rise with the casualty correctly positioned.

Note: Your other hand is free for use as needed.

> **WARNING**
>
> Do NOT use the neck drag if the casualty has a broken and/or fractured arm or a suspected neck injury. If the casualty is unconscious, protect his/her head from the ground.

    b. Neck drag.

        (1) Place the casualty on his back, if not already there. [See steps 3a (1)-(5)]. (

        (2) Tie the casualty's hands together at the wrists. (If conscious, the casualty may clasp his/her hands together around your neck.)

        (3) Straddle the casualty in a kneeling face-to-face position.

        (4) Loop the casualty's tied hands over and/or around your neck.

        (5) Crawl forward, looking ahead, dragging the casualty with you.

    c. Cradle-drop drag.

        (1) With the casualty lying on his/her back, kneel at the head.

        (2) Slide your hands, palms up, under the casualty's shoulders.

        (3) Get a firm hold under his/her armpits.

        (4) Partially rise, supporting the casualty's head on one of your forearms.

Note: You may bring your elbows together and let the casualty's head rest on both of your forearms.

        (5) With the casualty in a semisitting position, rise and drag the casualty backwards.

        (6) Back down the steps (or up if appropriate), supporting the casualty's head and body and letting the hips and legs drop from step to step.

    4. Evacuate the casualty using a SKED litter.

    a. Prepare the SKED litter for transport.

        (1) Remove the SKED from the pack and place on the ground.

        (2) Unfasten the retainer strap.

        (3) Step on the foot end of the SKED litter and unroll the SKED completely.

        (4) Bend the SKED in half and back roll.

        (5) Repeat with the opposite end of the litter so that the SKED litter lays flat.

        (6) Point out the handholds, straps for the casualty, and dragline at the head of the litter.

    b. Place and secure a casualty to a SKED litter.

        (1) Place the SKED litter next to the casualty so that the head end of the litter is next to the casualty's head.

        (2) Place the cross straps under the SKED litter.

        (3) Log roll the casualty onto his side in a steady and even manner.

        (4) Slide the SKED litter as far under the casualty as possible.

        (5) Gently roll the casualty until he is again lying on his back with the litter beneath him.

        (6) Slide the casualty to the middle of the SKED litter, keeping his spinal column as straight as possible.

        (7) Pull out the straps from under the SKED litter.

(8) Bring the straps across the casualty.

(9) Lift the sides of the SKED litter and fasten the four cross straps to the buckles directly opposite the straps.

(10) Lift the foot portion of the SKED litter.

(11) Feed the foot straps over the casualty's lower extremities and through the unused grommets at the foot end of the SKED litter.

(2) Fastens the straps to the buckles.

(13) Check to make sure the casualty is secured to the SKED litter.

c. Lift the casualty.

Note: For a SKED litter, lift the sides of the SKED and fasten the four cross straps to the buckles directly opposite the straps. Lift the foot portion of the SKED and feed the foot straps through the unused grommets at the foot end of the SKED and fasten to the buckles.

(1) Using four Soldiers (two on each side), all facing the casualty's feet. Have each rescuer grab a handle with their inside hand.

(2) In one fluid motion on the command of "prepare to lift, lift" raise as a unit holding the casualty parallel and even.

5. Evacuate the casualty using a Talon litter.

a. Prepare a Talon litter for use.

(1) Remove the litter from the bag.

(2) Stand the litter upright and release buckles from the litter.

(3) Place the litter on the ground and completely extend it with the fabric side facing up.

(4) Keeping the litter as straight as possible, grab the handles and rotate them inward until all the hinges rotate and lock.

Note: This action is done best using two individuals on each end of the litter executing this step simultaneously.

(5) While maintaining the hinges in the locked position, apply firm, steady pressure on the spreader bar with your foot. Increase pressure with your foot until the spreader bar locks into place.

b. Place the casualty on the litter.

(1) Place the litter next to the casualty. Ensure that the head end of the litter is beside the head of the casualty.

(2) Log roll the casualty and slide the litter as far under him/her as possible. Gently roll the casualty down onto the litter.

(3) Slide the casualty to the center of the litter. Be sure to keep the spinal column as straight as possible.

c. Secure the casualty to the litter using litter straps or other available materials.

6. Evacuate the casualty using an improvised litter.

a. Use the poncho and two poles or limbs.

(1) Open the poncho and lay the two poles lengthwise across the center, forming three equal sections.

(2) Reach in, pull the hood up toward you, and lay it flat on the poncho.

(3) Fold one section of the poncho over the first pole.

(4) Fold the remaining section of the poncho over the second pole to the first pole.

b. Use shirts or jackets and two poles or limbs.

(1) Zipper closed two uniform jackets and turn them inside out, leaving the sleeves inside.

(2) Lay the jackets on the ground and pass the poles through the sleeves, leaving one at the top and one at the bottom of the poles to support the casualty's whole body.

c. Place the casualty on the improvised litter.

(1) Lift the litter.

(2) Place the litter next to the casualty. Ensure the head end of the litter is adjacent to the head of the casualty.

(3) Slide the casualty to the center of the litter. Be sure to keep the spinal column as straight as possible.

(4) Secure the casualty to the litter using litter straps or other available materials.

7. Load casualties onto a military vehicle.

a. Ground ambulance.

Note: Ground ambulances have combat medics to take care of the casualties during evacuation. Follow any special instructions that they give for loading, securing, or unloading casualties.

(1) Make sure each litter casualty is secured to his litter. Use the litter straps when available.

(2) Load the most serious casualty last.

(3) Load the casualty head first (head in the direction of travel) rather than feet first.

(4) Make sure each litter is secured to the vehicle.

Note: Unload casualties in reverse order, most seriously injured casualty first.

b. Air ambulance.

Note: Air ambulances have combat medics to take care of the casualties during evacuation. Follow any special instructions that they give for loading, securing, or unloading casualties.

(1) Remain 50 yards from the helicopter until the litter squad is signaled to approach the aircraft.

WARNING

Never go around the rear of the UH-60 or UH-1 aircraft.

---

(2) Approach the aircraft in full view of the aircraft crew, maintaining visual confirmation that the crew is aware of the approach of the litter party. Ensure that the aircrew can continue to visually distinguish friendly from enemy personnel at all times. Maintain a low silhouette when approaching the aircraft.

(a) Approach UH-60/UH-1 aircraft from sides. Do not approach from the front or rear. If you must move to the opposite side of the aircraft, approach from the side to the skin of the aircraft. Then hug the skin of the aircraft, and move around the front of the aircraft to the other side.

(b) Approach CH-47/CH46 aircraft from the rear.

(c) Approach MH-53 aircraft from the sides to the rear ramp, avoiding the tail rotor.

(d) Approach nonstandard aircraft in full view of the crew, avoiding tail rotors, main rotors, and propellers.

(e) Approach high performance aircraft (M/C-130/-141B/-17/-5B) from the rear, under the guidance of the aircraft loadmaster or the ground control party.

(3) Load the most seriously injured casualty last.

(4) Load the casualty who will occupy the upper berth first, and then load the next litter casualty immediately under the first casualty.

Note: This is done to keep the casualty from accidentally falling on another casualty if his litter is dropped before it is secured.

(5) When casualties are placed lengthwise, position them with their heads toward the direction of travel.

(6) Make sure each litter casualty is secured to his litter

(7) Make sure each litter is secured to the aircraft.

Note: Unload casualties in reverse order, most seriously injured casualty first.

c. Ground military vehicles.

Note: Nonmedical military vehicles may be used to evacuate casualties when no medical evacuation vehicles area available.

Note: If medical personnel are present, follow their instructions for loading, securing, and unloading casualties.

(1) When loading casualties into the vehicle, load the most seriously injured casualty last.

(2) When a casualty is placed lengthwise, load the casualty with his head pointing forward, toward the direction of travel.

(3) Ensure each litter casualty is secured to the litter. Use litter straps, if available.

(4) Secure each litter to the vehicle as it is loaded into place. Make sure each litter is secured.

Note: Unload casualties in reverse order, most seriously injured casualty first

---

Evaluation Preparation:

Setup: For training and evaluation, use other Soldiers to be simulated casualties to be transported. Place Soldiers in both vehicles and on the ground for transport. Have at least one tactical vehicle available for loading, or at least a large platform area that can accommodate several litter casualties.

Brief Soldier: Tell the Soldier the simulated casualties require movement to the evacuation platform.

Performance Measures	GO	NO GO
1 Removed the casualty from a vehicle, if necessary.	_____	_____
2 Selected an appropriate method of transporting the casualty.	_____	_____
3 Evacuated the casualty using a manual carry.	_____	_____
4 Evacuated the casualty using a SKED litter.	_____	_____
5 Evacuated a casualty using a Talon litter.	_____	_____
6 Evacuated a casualty using an improvised litter(s).	_____	_____
7 Loaded casualties onto a military vehicle.	_____	_____

Evaluation Guidance: Score each Soldier according to the performance measures. Unless otherwise stated in the task summary, the Soldier must pass all performance measures to be scored GO. If the Soldier fails any steps, show the Soldier what was done wrong and how to do the task correctly.

Environment: Environmental protection is not just the law but the right thing to do. It is a continual process and starts with deliberate planning. Always be alert to ways to protect our environment during training and missions. In doing so, you will contribute to the sustainment of our training resources while protecting people and the environment from harmful effects. Refer to ATP 3-34.5 Environmental Considerations and GTA 05-08-002 ENVIRONMENTAL-RELATED RISK ASSESSMENT.

Safety: In a training environment, leaders must perform a risk assessment in accordance with ATP 5-19, Risk Management. Leaders will complete a DD Form 2977 DELIBERATE RISK ASSESSMENT WORKSHEET during the planning and completion of each task and sub-task by assessing mission, enemy, terrain and weather, troops and support available-time available and civil considerations, (METT-TC). *Note:* During MOPP training, leaders must ensure personnel are monitored for potential heat injury. Local policies and procedures must be followed during times of increased heat category in order to avoid heat related injury. Consider the MOPP work/rest cycles and water replacement guidelines IAW TM 3-11.32 Multi-Service Reference for Chemical, Biological, Radiological, and Nuclear Warning and Reporting and Hazard Prediction Procedures.

References:
Required:
Related: TC 4-02.1, ATP 4-25.13, ATP 4-02.2

081-COM-1007

Perform First Aid for Burns

Conditions: You have a casualty who has a burn injury. You will need the casualty's emergency bandage or field dressing and canteen of water. Some iterations of this task should be performed in MOPP 4.

Standards: Give first aid for a burn without causing further injury to the casualty.

Special Condition: None

Safety Risk: Low

MOPP: Sometimes

```
┌─────────────────────────────────────────────────────────┐
│                    Task Statements                       │
└─────────────────────────────────────────────────────────┘
```

Cue: None

Note: None

Performance Steps

1. Eliminate the source of the burn.

```
┌─────────────────────────────────────────────────────────┐
│                       CAUTION                            │
│  Synthetic materials, such as nylon, may melt and cause  │
│  further injury.                                         │
└─────────────────────────────────────────────────────────┘
```

a. Thermal burns. Remove the casualty from the source of the burn. If the casualty's clothing is on fire, cover the casualty with a field jacket or any large piece of nonsynthetic material and roll him/her on the ground to put out the flames.

WARNING

Do not touch the casualty or the electrical source with your bare hands. You will be injured too!

WARNING: High voltage electrical burns from an electrical source or lightning may cause temporary unconsciousness, difficulties in breathing, or difficulties with the heart (irregular heartbeat).

---

b. Electrical burns. If the casualty is in contact with an electrical source, turn the electricity off, if the switch is nearby. If the electricity cannot be turned off, use any nonconductive material (rope, clothing, or dry wood) to drag the casualty away from the source.

WARNING

Blisters caused by a blister agent are actually burns. Do not try to decontaminate skin where blisters have already formed. If blisters have not formed, decontaminate the skin.

---

c. Chemical burns.

(1) Remove liquid chemicals from the burned casualty by flushing with as much water as possible.

(2) Remove dry chemicals by carefully brushing them off with a clean, dry cloth. If large amounts of water are available, flush the area. Otherwise, do not apply water.

(3) Smother burning white phosphorus with water, a wet cloth, or wet mud. Keep the area covered with the wet material.

d. Laser burns. Move the casualty away from the source while avoiding eye contact with the beam source. If possible, wear appropriate laser eye protection.

Note: After the casualty has been removed from the source of the burn, continually monitor the casualty for conditions that may require basic lifesaving measures.

WARNING

Do NOT uncover the wound in a chemical environment. Exposure could cause additional harm.

---

2. Uncover the burn.

WARNING

Do NOT attempt to remove clothing that is stuck to the wound. Additional harm could result.

---

a. Cut clothing covering the burned area.

CAUTION

Do not pull clothing over the burns.

---

b. Gently lift away clothing covering the burned area.

c. If the casualty's hand(s) or wrist(s) have been burned, remove jewelry (rings, watches) and place them in his/her pockets.

3. Apply the casualty's dry, sterile dressing directly over the wound.

Note: If the burn is caused by white phosphorus, the dressing must be wet.

CAUTION:

Do not place the dressing over the face or genital area.

Do not break the blisters.

Do not apply grease or ointments to the burns.

a. Apply the dressing/pad, white side down, directly over the wound.

b. Wrap the tails (or the elastic bandage) so that the dressing/pad is covered.

c. For a field dressing, tie the tails into a nonslip knot over the outer edge of the dressing, not over the wound. For an emergency bandage, secure the hooking ends of the closure bar into the elastic bandage.

d. Check to ensure that the dressing is applied lightly over the burn but firmly enough to prevent slipping.

Note: If the casualty is conscious and not nauseated, give him/her small amounts of water to drink.

4. Watch the casualty closely for life-threatening conditions, check for other injuries (if necessary), and treat for shock. Seek medical aid.

5. Seek medical aid.

Evaluation Preparation:

Setup: For training and evaluation, use another Soldier to simulate a casualty with a burn injury.

Brief Soldier: Tell the Soldier to treat the casualty with a burn injury.

	Performance Measures	GO	NO GO
1	Eliminated the source of the burn.	_____	_____
2	Uncovered the burn, unless clothing was stuck to the wound or in a chemical environment.	_____	_____

Performance Measures	GO	NO GO
3 Applied the dressing/pad directly over the wound.	_____	_____

a. Applied the dressing/pad directly over the wound.
    b. Covered the edges of the dressing/pad.
    c. Properly secured the bandage.
    d. Applied the dressing lightly over the burn but firmly enough to prevent slipping.		

4 Watched the casualty closely for life-threatening conditions, checked for other injuries (if necessary), and treated for shock.	_____	_____
5 Sought medical aid.	_____	_____

Evaluation Guidance: Score each Soldier according to the performance measures. Unless otherwise stated in the task summary, the Soldier must pass all performance measures to be scored GO. If the Soldier fails any steps, show the Soldier what was done wrong and how to do the task correctly.

Environment: Environmental protection is not just the law but the right thing to do. It is a continual process and starts with deliberate planning. Always be alert to ways to protect our environment during training and missions. In doing so, you will contribute to the sustainment of our training resources while protecting people and the environment from harmful effects. Refer to ATP 3-34.5 Environmental Considerations and GTA 05-08-002 ENVIRONMENTAL-RELATED RISK ASSESSMENT. Environmental protection is not just the law but the right thing to do. It is a continual process and starts with deliberate planning. Always be alert to ways to protect our environment during training and missions. In doing so, you will contribute to the sustainment of our training resources while protecting people and the environment from harmful effects. Refer to ATP 3-34.5 Environmental Considerations and GTA 05-08-002 ENVIRONMENTAL-RELATED RISK ASSESSMENT.

Safety: In a training environment, leaders must perform a risk assessment in accordance with ATP 5-19, Risk Management. Leaders will complete a DD Form 2977 DELIBERATE RISK ASSESSMENT WORKSHEET during the planning and completion of each task and sub-task by assessing mission, enemy, terrain and weather, troops and support available-time available and civil considerations, (METT-TC). *Note:* During MOPP training, leaders must ensure personnel are monitored for potential heat injury. Local policies and procedures must be followed during times of increased heat category in order to avoid heat related injury. Consider the MOPP work/rest cycles and water replacement guidelines IAW TM 3-11.32 Multi-Service Reference for Chemical, Biological,

Radiological, and Nuclear Warning and Reporting and Hazard Prediction Procedures.

081-COM-0101

Request Medical Evacuation

Conditions: You have a casualty requiring medical evacuation (MEDEVAC). All medical interventions have been completed and the casualty is stable. This task should not be trained in MOPP 4

Standards: Transmit a 9-Line MEDEVAC request, providing all necessary information as quickly as possible in accordance with (IAW) Army Training Publication (ATP) 4-02.2, Medical Evacuation

Safety Risk: Low

MOPP 4: Sometimes

Task Statements

Cue: None

Note: For non-APD references contact your training NCO and or check with the MOS library.

Performance Steps

1. Collect all applicable information needed for the MEDEVAC request.

    a. Determine the grid coordinates for the pickup site. (See STP 21-1-SMCT, task 071-COM-1002.)

    b. Obtain radio frequency, call sign, and suffix.

    c. Obtain the number of patients and precedence.

    d. Determine the type of special equipment required.

    e. Determine the number and type (litter or ambulatory) of patients.

    f. Determine the security of the pickup site.

    g. Determine how the pickup site will be marked.

    h. Determine patient nationality and status.

    i. Obtain pickup site chemical, biological, radiological, and nuclear (CBRN) contamination information normally obtained from the senior person or medic.

Note: CBRN line 9 information is only included when contamination exists.

2. Record the gathered MEDEVAC information using the authorized brevity codes. (See tables 081-COM-0101-1 and 081-COM-0101-2.)

Note: Unless the MEDEVAC information is transmitted over secure communication systems, it must be encrypted, except as noted in step 3b(1).

LINE	ITEM	EXPLANATION	WHERE/HOW OBTAINED	WHO NORMALLY PROVIDES	REASON
1	Location of pickup site	Encrypt the grid coordinates of the pickup site. When using the DRYAD Numeral Cipher, the same "SET" line will be used to encrypt the grid zone letters and the coordinates. To preclude misunderstanding, a statement is made that grid zone letters are included in the message (unless unit SOP specifies its use at all times).	From map	Unit leader(s)	Required so evacuation vehicle knows where to pick up patient. Also, so that the unit coordinating the evacuation mission can plan the route for the evacuation vehicle (if the evacuation vehicle must pick up from more than one location).
2	Radio frequency, call sign, and suffix	Encrypt the frequency of the radio at the pickup site, not a relay frequency. The call sign (and suffix if used) of person to be contacted at the pickup site may be transmitted in the clear.	From SOI	RTO	Required so that evacuation vehicle can contact requesting unit while en route (obtain additional information or change in situation or directions).
3	Number of patients by precedence	Report only applicable information and encrypt the brevity codes. A - URGENT B - URGENT-SURG C - PRIORITY D - ROUTINE E - CONVENIENCE If two or more categories must be reported in the same request, insert the word "BREAK" between each category.	From evaluation of patient(s)	Medic or senior person present	Required by unit controlling vehicles to assist in prioritizing missions.
4	Special equipment required	Encrypt the applicable brevity codes. A - None B - Hoist C - Extraction equipment D - Ventilator	From evaluation of patient/situation	Medic or senior person present	Required so that the equipment can be placed on board the evacuation vehicle prior to the start of the mission.
5	Number of patients by type	Report only applicable information and encrypt the brevity code. If requesting medical evacuation for both types, insert the word "BREAK" between the litter entry and ambulatory entry. L + # of patients - Litter A + # of patients - Ambulatory (sitting)	From evaluation of patient(s)	Medic or senior person present	Required so that the appropriate number of evacuation vehicles may be dispatched to the pickup site. They should be configured to carry the patients requiring evacuation.
6	Security of pickup site (wartime)	N - No enemy troops in area P - Possibly enemy troops in area (approach with caution) E - Enemy troops in area (approach with caution) X - Enemy troops in area (armed escort required)	From evaluation of situation	Unit leader	Required to assist the evacuation crew in assessing the situation and determining if assistance is required. More definitive guidance can be furnished the evacuation vehicle while it is en route (specific location of enemy to assist an aircraft in planning its approach).

Table 081-COM-0101-1
Lines 1-6

Performance Steps

LINE	ITEM	EXPLANATION	WHERE/HOW OBTAINED	WHO NORMALLY PROVIDES	REASON
6	Number and type of wound, injury, or illness (peacetime)	Specific information regarding patient wounds by type (gunshot or shrapnel). Report serious bleeding, along with patient's blood type, if known.	From evaluation of patient(s)	Medic or senior person present	Required to assist evacuation personnel in determining treatment and special equipment needed.
7	Method of marking pickup site	Encrypt the brevity codes. A - Panels B - Pyrotechnic signal C - Smoke signal D - None E - Other	Based on situation and availability of materials	Medic or senior person present	Required to assist the evacuation crew in identifying the specific location of the pickup. Note that the color of the panels or smoke should not be transmitted until the evacuation vehicle contacts the unit (just prior to its arrival). For security, the crew should identify the color and the unit verifies it.
8	Patient nationality and status	The number of patients in each category need not be transmitted. Encrypt only the applicable brevity codes. A - US military B - US citizen C - Non-US military D - Non-US citizen E - Enemy prisoner of war (EPW)	From evaluation of patient(s)	Medic or senior person present	Required to assist in planning for destination facilities and need for guards. Unit requesting support should ensure that there is an English-speaking representative at the pickup site.
9	CBRN contamination (wartime)	Include this line only when applicable. Encrypt the applicable brevity codes. C - Chemical B - Biological R - Radiological N - Nuclear	From situation	Medic or senior person present	Required to assist in planning for the mission (determine which evacuation vehicle will accomplish the mission and when it will be accomplished).
9	Terrain description (peacetime)	Include details of terrain features in and around proposed landing site. If possible, describe relationship of a site to prominent terrain feature (lake, mountain, tower).	From area survey	Personnel present	Required to allow evacuation personnel to assess route/avenue of approach into area. Of particular importance if hoist operation is required.

Table 081-COM-0101-2
Lines 6-9

a. Location of the pickup site (line 1).

b. Radio frequency, call sign, and suffix (line 2).

c. Numbers of patients by precedence (line 3).

(1) Encrypt this information using the following brevity codes:
A=Urgent. B= Urgent Surgical. C= Priority. D= Routine. E= Convenience.

(2) If 2 or more categories are reported in same request, insert the word "break" between each category.

d. Special equipment required (line 4). Encrypt this information using the following brevity codes: A= None. B= Hoist. C= Extraction Equipment. D= Ventilator.

e. Number of patients by type (line 5). Encrypt this information using the following brevity codes: L+#: Number of litter patients. A+#: Number of ambulatory patients (able to walk or can walk with assistance).
Note: If requesting MEDEVAC for both types, insert the word "break" between the litter entry and the ambulatory entry.

f. Security of the pickup site (line 6- wartime). Encrypt this information using the following brevity codes: N= No enemy troops in area. P= Possibly enemy troops in area, approach with caution. E= Enemy troops in area, approach with caution. X= Enemy troops in area, armed escort required.

g. Number and type of wound, injury or illness (line 6- peacetime)

h. Method of marking the pickup site (line 7). Encrypt this information using the following brevity codes: A= Panels. B= Pyrotechnic signal. C= Smoke signal. D= None. E= Other.

i. Patient nationality and status (line 8). Encrypt this information using the following brevity codes: A= US Military. B= US Civilian. C= Non-US Military. D= Non-US Civilian. E= Enemy prisoner (EPW).

j. CBRN contamination (line 9). Encrypt this information using the following brevity codes: N= Nuclear or radiological. B= Biological. C= Chemical.

k. Terrain Description (line 9 - peacetime)

3. Transmit the MEDEVAC request. (See STP 21-1-SMCT, task 113-COM-1022.)

Note: Transmission may vary depending on individual experience level and situation.

a. Contact the unit that controls the evacuation assets.

(1) Make proper contact with the intended receiver. Use effective call sign and frequency assignments from the SOI.

(2) Give the following in the clear "I HAVE A MEDEVAC REQUEST;" wait one to three seconds for a response. If no response, repeat the statement.

b. Transmit the MEDEVAC information in the proper sequence.

(1) State all line item numbers in clear text. The call sign and suffix (if needed) in line 2 may be transmitted in the clear text.

Note: Line numbers 1 through 5 must always be transmitted during the initial contact with the evacuation unit. Lines 6 through 9 may be transmitted while the aircraft or vehicle is en route.

(2) Follow the procedure provided in the explanation column of the MEDEVAC request format to transmit other required information. (See tables 081-COM-0101-1 and 081-COM-0101-2.)

(3) Pronounce letters and numbers according to appropriate radiotelephone procedures.

(4) End the transmission by stating "OVER."

(5) Keep the radio on and listen for additional instructions or contact from the evacuation unit.

4. Keep the radio on and listen for additional instructions or contact from the evacuation unit.

(Asterisks indicates a leader performance step.)

Evaluation Guidance: Score each Soldier according to the performance measures in the evaluation guide. Unless otherwise stated in the task summary, the Soldier must pass all performance measures to be scored GO. If the Soldier fails any step, show what was done wrong and how to do it correctly.

Evaluation Preparation: You must evaluate the students on their performance of this task in a field condition related to the actual task.

Performance Measures	GO	NO GO
1 Collected all information needed for the MEDEVAC request line items 1 through 9	____	____

Performance Measures	GO	NO GO
Note: Wartime procedures for line items 6 and 9 will be used.		
2 Recorded the gathered MEDEVAC information using the authorized brevity codes.	_____	_____
3 Transmitted the MEDEVAC request as quickly as possible, following appropriate radiotelephone procedures.	_____	_____
4 Kept the radio on, listened for additional instruction or contact from the evacuation unit.	_____	_____

References: ATP 4-02.2; ISBN 9781284041750; ATP 6-02.53

Environment: Environmental protection is not just the law but the right thing to do. It is a continual process and starts with deliberate planning. Always be alert to ways to protect our environment during training and missions. In doing so, you will contribute to the sustainment of our training resources while protecting people and the environment from harmful effects. Refer to the current Environmental Considerations manual and the current GTA Environmental-related Risk Assessment card.

Safety: In a training environment, leaders must perform a risk assessment in accordance with current Risk Management Doctrine. Leaders will complete the current Deliberate Risk Assessment Worksheet in accordance with the TRADOC Safety Officer during the planning and completion of each task and sub-task by assessing mission, enemy, terrain and weather, troops and support available-time available and civil considerations, (METT-TC). *Note*: During MOPP training, leaders must ensure personnel are monitored for potential heat injury. Local policies and procedures must be followed during times of increased heat category in order to avoid heat related injury. Consider the MOPP work/rest cycles and water replacement guidelines IAW current CBRN doctrine.

Prerequisite Individual Tasks : None

052-COM-1271

Identify visual Indicators of an Implosive Device (IED) **(Located at**
https://www.us.army.mil/suite/doc/23838510**)**
(UNCLASSIFIED//FOR OFFICIAL USE ONLY) (U//FOUO)

Conditions: This task is identified as FOUO, refer to DTMS or CAR to view

052-COM-1270

React to an Improvised Explosive Device (IED) Attack **(Located at**
https://www.us.army.mil/suite/doc/23838478**)**
(UNCLASSIFIED//FOR OFFICIAL USE ONLY) (U//FOUO)

Conditions: This task is identified as FOUO, refer to DTMS or CAR to view

071-COM-0815

Practice Noise, Light, and Litter Discipline

Conditions: You are member of a mounted or dismounted element conducting a
tactical mission and have been directed to comply with noise, light and litter
discipline. Enemy elements are in your area of operation. Some iterations of this
task should be performed in MOPP 4.

Standards: Prevent enemy from locating your element by exercising noise, light,
and litter discipline at all times.

Special Condition: None

Safety Risk: Low

MOPP 4: Sometimes

Task Statements

Cue: None

*Note:*None

Performance Steps

1. Exercise noise discipline.

   a. Avoid all unnecessary vehicular and foot movement.

   b. Secure (with tape or other materials) metal parts (for example, weapon slings, canteen cups, identification [ID] tags) to prevent them from making noise during movement.

Note: Do not obstruct the moving parts of weapons or vehicles.

   c. Avoid all unnecessary talk.

   d. Use radio only when necessary.

   e. Set radio volume low so that only you can hear.

   f. Use visual techniques to communicate.

2. Exercise light discipline.

   a. Do not smoke.

Note: The smoking of cigarettes, cigars, etc., can be seen and smelled by the enemy.

   b. Conceal flashlights and other light sources so that the light is filtered (for example, under a poncho).

   c. Cover or blacken anything that reflects light (for example, metal surfaces, vehicles, glass).

   d. Conceal vehicles and equipment with available natural camouflage.

3. Exercise litter discipline.

   a. Establish a litter collection point (empty food containers, empty ammunition cans or boxes, old camouflage) when occupying a position.

   b. Verify all litter has been collected in preparation to leaving a position.

   c. Take all litter with you when leaving a position.

Evaluation Preparation:

Setup: Provide the Soldier with the equipment and or materials described in the conditions statement.

Brief Soldier: Tell the Soldier what is expected of him by reviewing the task standards. Stress to the Soldier the importance of observing all cautions, warnings, and dangers to avoid injury to personnel and, if applicable, damage to equipment.

Performance Measures	GO	NO GO
1 Exercised noise discipline.	_____	_____
2 Exercised light discipline.	_____	_____
3 Exercised litter discipline.	_____	_____

Evaluation Guidance: Score the Soldier GO if all performance measures are passed. Score the Soldier NO-GO if any performance measure is failed. If the Soldier scores a NO-GO, show the Soldier what was done wrong and how to do it correctly.

References
Required:
Related: TC 3-21.75

071-COM-0804

Perform Surveillance without the Aid of Electronic Device

Conditions: You are a member of a squad or team in a defensive position and must conduct surveillance within your assigned sector during both daylight and limited visibility (night). Some iterations of this task should be performed in MOPP 4.

Standards: Identify potential activity indicators and conduct a visual search of your assigned sector. Submit SALUTE reports, as required.

Special Condition: None

Safety Risk: Low

MOPP 4: Sometimes

Task Statements

Cue: None

Note: None

Performance Steps

1. Identify potential activity indicators in sector (Figure 071-COM-0804-1).

SIGHT Look for--	SOUND Listen for--	TOUCH Feel for--	SMELL Smell for--
• Enemy personnel, vehicles, and aircraft • Sudden or unusual movement • New local inhabitants • Smoke or dust • Unusual movement of farm or wild animals • Unusual activity--or lack of activity--by local inhabitants, especially at times or places that are normally inactive or active	• Running engines or track sounds • Voices • Metallic sounds • Gunfire, by weapon type • Unusual calm or silence • Dismounted movement • Aircraft	• Warm coals and other materials in a fire • Fresh tracks • Age of food or trash	• Vehicle exhaust • Burning petroleum products • Food cooking • Aged food in trash • Human waste

OTHER CONSIDERATIONS	
Armed Elements	Locations of factional forces, mine fields, and potential threats.
Homes and Buildings	Condition of roofs, doors, windows, lights, power lines, water, sanitation, roads, bridges, crops, and livestock.
Infrastructure	Functioning stores, service stations, and so on.
People	Numbers, gender, age, residence or DPRE status, apparent health, clothing, daily activities, and leadership.
Contrast	Has anything changed? For example, are there new locks on buildings? Are windows boarded up or previously boarded up windows now open, indicating a change in how a building is expected to be used? Have buildings been defaced with graffiti?

Additional SIGHT items (continued):
• Vehicle or personnel tracks
• Movement of local inhabitants along uncleared routes, areas, or paths
• Signs that the enemy has occupied the area
• Evidence of changing trends in threats
• Recently cut foliage
• Muzzle flashes, lights, fires, or reflections
• Unusual amount (too much or too little) of trash

Figure 071-COM-0804-1. Potential Indicators.

2. Perform observation techniques of the sector.
    a. Conduct day observation.
        (1) Use rapid scan technique. (Figure 071-COM-0804-2).
Note: The rapid scan technique is used to detect obvious signs of enemy activity. It is usually the first method you will use.

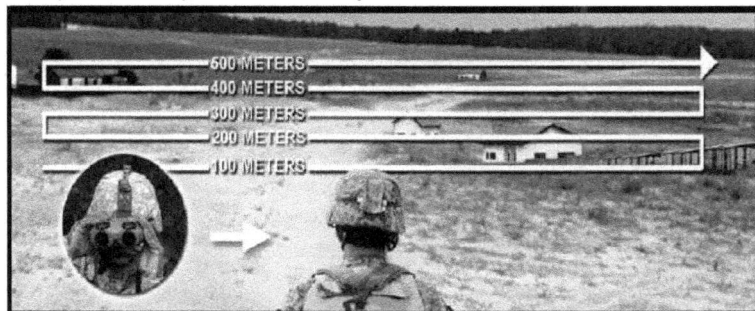

Figure 071-COM-0804-2. Rapid/Slow Scan.

(a) Search a strip of terrain about 100 meters deep, from left-to-right, pausing at short intervals.

(b) Search another 100-meter strip farther out, from right-to-left, overlapping the first strip scanned, pausing at short intervals.

(c) Continue this method until the entire sector of fire has been searched.

(2) Use slow scan technique.

Note: Slow scan search technique uses the same process as the rapid scan but much more deliberately; this means a slower, side-to-side movement and more frequent pauses.

(3) Use detailed search technique paying attention to the following: (Figure 071-COM-0804-3).

Note: The detailed search, even more than the rapid or slow scan, depends on breaking a larger sector into smaller sectors to ensure everything is covered in detail and no possible enemy positions are overlooked.

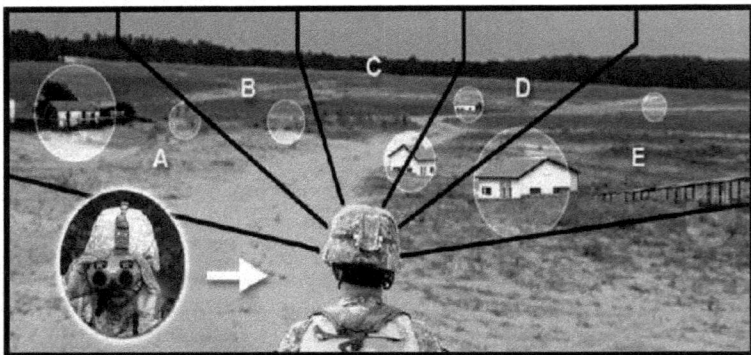

Figure 071-COM-0804-3. Detailed Search.

(a) Likely enemy positions and suspected vehicle/dismounted avenues of approach.

(b) Target signatures, such as road junctions, hills, and lone buildings, located near prominent terrain features.

(c) Areas with cover and concealment, such as tree lines and draws.

b. Conduct limited visibility observation.

(1) Use dark adaptation technique.

(a) Stay in a dark area for about 30 minutes.

(b) Move into a red-light area for about 20 minutes followed by about 10 minutes in a dark area.

Note: The red-light method may save time by allowing you to get orders, check equipment, or do some other job before moving into darkness.

(2) Use night vision scan technique (Figure 071-COM-0804-4).

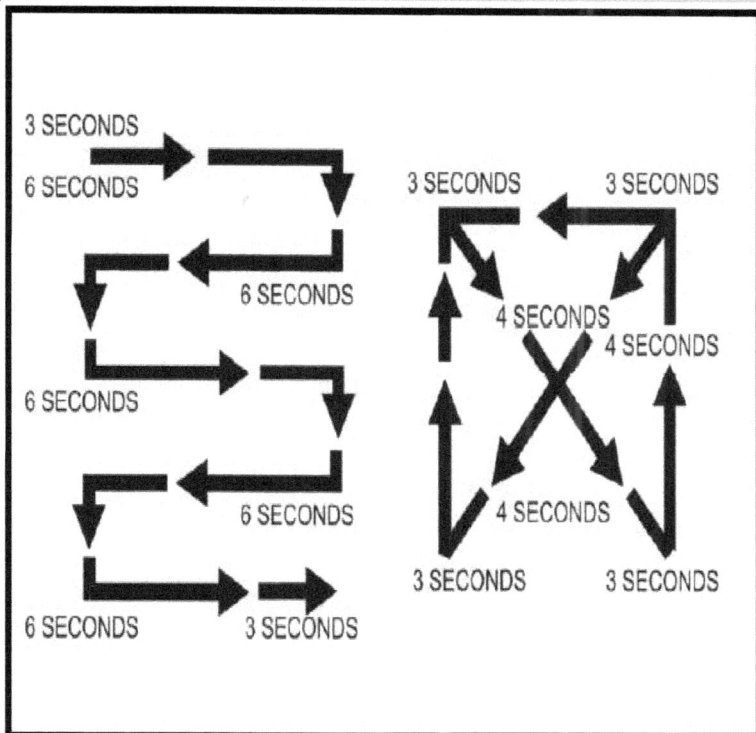

Figure 071-COM-0804-4. Night scanning patterns.

       (a) Look from right to left or left to right using a slow, regular scanning movement.

       (b) At night avoid looking directly at a faintly visible object when trying to confirm its presence.

       (3) Use off center vision technique.

Note: The technique of viewing an object using central vision is ineffective at night due to the night blind spot that exist during low illumination. You must learn to use off-center vision.

       (a) View an object by looking 10 degrees above, below, or to either side of it rather than directly at it.

       (b) Shift your eyes from one off-center point to another.

       (c) Continue to pick-up the object in your peripheral field of vision.

    3. Submit SALUTE report (Figure 071-COM-0804-5).

Performance Steps

Line No.	Type Info	Description
1	(S)ize/Who	Expressed as a quantity and echelon or size. For example, report "10 enemy Infantrymen" (not "a rifle squad").
	If multiple units are involved in the activity you are reporting, you can make multiple entries.	
2	(A)ctivity/What	Relate this line to the PIR being reported. Make it a concise bullet statement. Report what you saw the enemy doing, for example, "emplacing mines in the road."
3	(L)ocation/Where	This is generally a grid coordinate, and should include the 100,000-meter grid zone designator. The entry can also be an address, if appropriate, but still should include an eight-digit grid coordinate. If the reported activity involves movement, for example, advance or withdrawal, then the entry for location will include "from" and "to" entries. The route used goes under "Equipment/How."
4	(U)nit/Who	Identify who is performing the activity described in the "Activity/What" entry. Include the complete designation of a military unit, and give the name and other identifying information or features of civilians or insurgent groups.
5	(T)ime/When	For future events, give the DTG for when the activity will initiate. Report ongoing events as such. Report the time you saw the enemy activity, not the time you report it. Always report local or Zulu (Z) time.
6	(E)quipment/How	Clarify, complete, and expand on previous entries. Include information about equipment involved, tactics used, and any other essential elements of information (EEI) not already reported in the previous lines.

Figure 071-COM-0804-5. SALUTE Format.

Evaluation Preparation:

Setup: Provide the Soldier with the equipment and or materials described in the conditions statement.

Brief Soldier: Tell the Soldier what is expected of him by reviewing the task standards. Stress to the Soldier the importance of observing all cautions, warnings, and dangers to avoid injury to personnel and, if applicable, damage to equipment.

Performance Measures	GO	NO GO
1 Identified potential activity indicators in sector.	_____	_____

Performance Measures	GO	NO GO
2 Performed observation techniques of the sector.	____	____
3 Submitted SALUTE report.	____	____

Evaluation Guidance: Score the Soldier GO if all performance measures are passed. Score the Soldier NO-GO if any performance measure is failed. If the Soldier scores a NO-GO, show the Soldier what was done wrong and how to do it correctly.

References:
Required:
Related: TC 3-21.75

071-COM-0801

Challenge Persons Entering Your Area

Conditions: You are a member of a squad or team providing security for your unit in a field environment. You have your assigned weapon, individual protective equipment, and the current challenge and password. An unknown person or persons is approaching your area. Some iterations of this task should be performed in MOPP 4.

Standards: Detect and challenge all approaching personnel; prevent compromise of password; allow personnel positively identified as friendly to pass; and disarm, detain and report personnel not positively identified.

Special Condition: None

Safety Risk: Low

MOPP 4: Sometimes

Task Statements

Cue: None

Note: None

1. Detect all personnel entering your area.
2. Challenge an individual that enters your area.

    a. Cover the individual with your weapon without disclosing your position.

    b. Command the individual to "HALT" before they are close enough to pose a threat.

Note: Commands and questions must be loud enough to be heard by the individual but not loud enough that others outside of the immediate area can hear. Commands should be repeated as necessary.

    c. Ask "WHO IS THERE?" just loud enough for the individual to hear.

Note: The individual should reply with an answer that best describes them, example "Sergeant Jones".

    d. Order the individual to "ADVANCE TO BE RECOGNIZED".

    e. Continue to keep individual covered without exposing yourself.

    f. Order the individual to "HALT" when they are is within 2 to 3 meters from your position.

Note: The individual should be halted at a location that provides protection to you and prevents them from escaping if they are deemed unfriendly.

    g. Issue the challenge in a low voice.

Note: The challenge should only be heard by the individual challenged to prevent all others from overhearing. You may also ask the individual questions that only a friendly person should be able answer correctly.

    h. Determine if the individual is friendly based upon their return of the correct password and your own situational awareness.

    (1) Allow the individual to pass if the individual returns the correct password and you are convinced the individual is friendly.

    (2) Detain an individual if they return an incorrect password or cannot be positively identified as friendly.

    (a) Direct the individual to disarm.

    (b) Notify your chain of command.

    (c) Await instruction from your command.

3. Challenge a group that enters your area.

Note: These procedure and precautions are similar to those for challenging a single person.

    a. Cover the group with your weapon without disclosing your position.

    b. Order the group to halt before they are close enough to pose a threat to you.

    c. Command "WHO IS THERE?" just loud enough to be heard by the group.

    d. Wait for reply from group.

Note: Reply should clearly identify group, example "Friendly Patrol".

    e. Order the leader of the group to "ADVANCE TO BE RECOGNIZED".

    f. Continue to keep group leader covered without exposing yourself.

    g. Order the group leader to "HALT" when the individual is within 2 to 3 meters from your position.

Note: The group leader should be halted at a location that provides protection to you and prevents the leader from escaping if they are deemed unfriendly.

h. Issue the challenge to only the group leader.

Note: You may also ask questions that only a friendly person should be able to answer correctly.

i. Determine if the group leader is friendly based upon their return of the correct password and your own situational awareness.

(1) Direct the group leader to remain with you to assist in identifying group members, if you determine he/she is friendly.

(a) Direct the group leader to vouch for or positively identify each member of his group as they pass to your flank.

(b) Detain any individual in the group not recognized by the group leader by disarming them, and having them wait until your chain of command provides additional instructions.

(2) Detain the group leader, if not positively identified as friendly.

(a) Direct the individual to disarm.

(b) Direct him/her to inform their group to wait.

(c) Notify your chain of command.

(d) Await instruction from your chain of command.

Evaluation Preparation:

Setup: Provide the Soldier with the equipment and or materials described in the conditions statement.

Brief Soldier: Tell the Soldier what is expected of him by reviewing the task standards. Stress to the Soldier the importance of observing all cautions, warnings, and dangers to avoid injury to personnel and, if applicable, damage to equipment.

Performance Measures	GO	NO GO
1 Detected all personnel entering your area.	___	___
2 Challenged all individuals that entered your area.	___	___
3 Challenged all groups that entered your area.	___	___

Evaluation Guidance: Score the Soldier GO if all performance measures are passed. Score the Soldier NO-GO if any performance measure is failed. If the Soldier scores a NO-GO, show the Soldier what was done wrong and how to do it correctly.

References:
Required:

301-COM-1050

Report Information of Potential Intelligence Value

<table>
<tr><th>WARNING</th></tr>
<tr><td>Do not wait until you have complete information to transmit. Even small amounts of information of critical tactical value may provide indicators of the threat's intentions.</td></tr>
</table>

Conditions: You are a Soldier with the responsibility to actively observe and provide concise accurate reports while in an area of operations. You are given information requirements, a means of communication (radio, wire, cable, or messenger) as prescribed in the unit's standard operating procedures (SOPs), required mission-specific equipment and a situation which requires you to immediately report information of critical tactical value.

Standards: Transmit information to the receiving authority in size, activity, location, unit, time and equipment (SALUTE) format to include significant terrain and weather conditions via the available means of communication. Information will be reported within 5 minutes after observation with six out of six SALUTE items correctly identified. *Note:* Your unit SOPs will specify the receiving authority. Examples of receiving authorities are company commander, team commander, company intelligence support team (CoIST), or S2 (Intelligence Officer [U.S. Army]) section.

Special Condition: None

Safety Risk: low

MOPP 4:

Task Statements

Cue: None

*Note:*None

1. Identify information concerning threat activity and significant terrain and weather conditions including-

a. Order of battle factors; for example, threat weapons systems, composition, and direction of movement.

Note: If you cannot identify a weapon system or vehicle by name, include a description of the equipment.

b. Military aspects of terrain; for example, observation and fields of fire, avenues of approach, key and decisive terrain, obstacles, and cover and concealment (OAKOC).

c. Weather factors; for example, severe weather, precipitation, trafficability, surface winds and gusts, and ground visibility.

Note: Use Spot Reports (Level 1 Report) to transmit information of immediate value. Transmit Spot Reports as rapidly and securely as possible. The SALUTE format is an aid for the observer to report the essential reporting elements. (You may precede each message segment of the Spot Report with the meaning of the acronym SALUTE.)

2. Draft message summary information in the SALUTE format.

a. S-Size. Report the number of personnel, vehicles, aircraft, or size of an object. Make an estimate if necessary.

b. A-Activity. Report detailed account of the detected element activity. Indicate the activity types or types and an amplifying sub-type if applicable.

(1) Attacking. (direction from)
(a) ADA. (engaging)
(b) Aircraft. (engaging) (rotary wing (RW), fixed wing (FW))
(c) Ambush. (improvised explosive device (IED) (exploded), IED (unexploded), Sniper, Anti-armor, Other)
(d) Indirect fire. (point of impact, point of origin)
(e) Chemical, biological, radiological, and nuclear (CBRN)
(2) Defending. (direction from)
(3) Moving. (direction from)
(4) Stationary.
(5) Cache.
(6) Civilian. (criminal acts, unrest, infrastructure damage)
(7) Personnel recovery. (isolating event, observed signal)
(8) Other. (Give name and description)

c. L-Location. Report where you saw the activity. Include grid coordinates with Grid Zone Designator or reference from a known point including the distance and direction from the known point.

d. U-Unit. Report the detected element unit, organization, or facility. Indicate the type of unit, organization, or facility detected. If it cannot be clearly identified, describe in as much detail as possible, including uniforms, vehicle markings, and other identifying information.

(1) Conventional.
(2) Irregular.
(3) Coalition.
(4) Host Nation.
(5) Non-governmental Organization (NGO).

(6) Civilian.

(7) Facility.

e. T-Time. Report the time and date the activity was observed, not the time you report it. Always report local or Zulu time.

f. E-Equipment. Report all equipment associated with the activity, such as weapons, vehicles, tools. Add a narrative if necessary to clarify, describe, or explain the type of equipment. Provide nomenclature, type, and quantity of all equipment observed. If equipment cannot be clearly identified, describe in as much detail as possible.

(1) Air Defense Artillery (ADA) (missile (MANPADS), missile (other), gun)

(2) Artillery (gun (self propelled), gun (towed), missile or rocket, mortar)

(3) Armored track vehicle (tank, APC, command and control (C2), engineer, transport, other)

(4) Armored wheel vehicle (gun, APC, C2, engineer, transport, other)

(5) Wheel vehicle (gun, C2, engineer, transport, other)

(6) Infantry weapon (anti-armor missile, anti-armor gun, RPG, heavy machinegun, GL, small arms, other)

(7) Aircraft (RW (attack helicopter (AH)), RW (utility helicopter (UH)), RW (observation helicopter), FW (attack), FW (transport), unmanned aircraft, other)

(8) Mine or IED (buried, surface, VBIED, PBIED, other)

(9) CBRN

(10) Supplies (Class III, Class V, other)

(11) Civilian

(12) Other

3. Select a means of communication; for example, radio, wire, cable, or messenger.

Note: Consider the communications means available and the information's potential significance to your mission. Radio is fast and mobile; yet, normally it is the least secure of the three communications means available at tactical units. Wire is more secure but it is subject to wiretapping and requires more time, personnel and equipment to install. Messenger is very secure but requires more delivery time and is limited by weather, terrain, and threat action.

4. Transmit the message to the receiving authority.

a. If using a messenger, provide the messenger with explicit reporting instructions and a message, preferably written, which is clear, complete, and concise.

b. If using radio, use proper radio/telephone procedures according to unit SOPs. Use the radio only as needed. The enemy may intercept your transmission, exploit the message information, or locate your transmitter for targeting or jamming.

c. If you encounter jamming or interference on your radio net, within 10 minutes of the incident, transmit a meaconing, intrusion, jamming, and interference (MIJI) feeder report, preferably via messenger, wire, or cable to your net control station. Your Signal Operating Instructions (SOI) contains the MIJI format.

Evaluation Preparation:

Setup: Simulate a situation that requires Soldiers to immediately report information of critical tactical value. You may need two to four personnel (dressed in aggressor uniforms or local attire if available) where they are observable with the naked eye (or binoculars if available). Direct the personnel to perform some type of activity that meets the information requirements. Provide the Soldier with a 1:50,000 scale topographic map of the test area. Provide paper and a pen or pencil for the Soldier to take notes and prepare the report. If you require the Soldier to radio the report to someone else, provide two radios and SOI. Accompany the Soldier being tested to a location where the Soldier can observe the threat.

Brief Soldier: Tell the Soldier he/she is–
Performing an offensive or defensive mission.
Patrolling in a stability or defense support of civil authorities operation.
Manning a checkpoint or roadblock.
Occupying an observation post.
Passing through an area in a convoy.
Instruct the Soldier to report the activity observed, weather factors, and any significant military aspects of the terrain. Once the Soldier completes the report, have the Soldier select a means of transmitting the report to the receiving authority.

Performance Measures	GO	NO GO
1 Identified	____	____
a. Order of battle factors; for example, threat weapons systems, composition, and direction of movement.		
b. Military aspects of terrain; for example, OAKOC.		
c. Weather factors; for example, severe weather, precipitation, trafficability, surface winds and gusts, and ground visibility.		
2 Drafted a message in SALUTE format identifying	____	____
a. Size.		
b. Activity.		
c. Location.		

Performance Measures	GO	NO GO
d. Unit.		
e. Time.		
f. Equipment		
3 Selected a means of communication.	_____	_____
4 Transmitted the message to the receiving authority within 5 minutes of the observation.	_____	_____

Evaluation Guidance: Refer to chapter 1, paragraph 1-9e, (1) and (2).

References:
Required: TC 3-21.75, FM 6-99
Related: ATP 3-55.4

071-COM-1004

Perform Duty as a Guard

Conditions: You are a Soldier and have been assigned to a guard post. The Sergeant of the Guard has given you any guard post-specific equipment, special orders, and local standard operating procedures (SOP). You have your individual weapon and personal equipment.

Standards: Perform duty on assigned guard post in accordance with special orders and SOP.

Special Condition: None
Safety Risk: Low
MOPP 4:

Task Statements

Performance Step

1. Prepare for guard duty.
    a. Review general orders, special orders, and local implementing SOPs.
    b. Inspect all required equipment for serviceability.
2. Assume guard post duty.

a. Receive special order changes, if any.

b. Establish communications with the Relief Commander or the Sergeant of the Guard via FM or telephonic means.

Note: During a duty tour a guard is required to execute orders ONLY from the commanding officer, the field officer of the day, the officer of the day and officers of the guard.

3. Walk the guard post or assume the guard position.

Note: While on guard duty, surrender your weapon to, and only on order of, a person from whom you lawfully receives orders while on post.

4. Challenge personnel as specified by the special orders. *Note:* Challenge position is port arms or raised pistol.

a. Challenge all suspicious individuals observed.

b. Challenge a group as specified by the special orders.

c. Salute officers when performing guard duty on guard posts that do not require a challenge.

5. Pass instructions and changes to orders on to relief guard.

(Asterisks indicates a leader performance step.)

Evaluation Guidance: Score the Soldier GO if all performance measures are passed. Score the Soldier NO-GO if any performance measure is failed. If the Soldier scores a NO-GO, show the Soldier what was done wrong and how to do it correctly.

Evaluation Preparation: SETUP: Provide the Soldier with the equipment and/or materials described in the conditions statement.

Brief Soldier: Tell the Soldier what is expected by reviewing the task standards. Stress to the Soldier the importance of observing all cautions, warnings, and dangers to avoid injury to personnel and, if applicable, damage to equipment.

	Performance Measures	GO	NO GO
1	Prepare for guard duty.	_____	_____
2	Assumed guard duty.	_____	_____
3	Walked guard post or assumed guard position.	_____	_____
4	Challenged personnel as specified by the special orders.	_____	_____
5	Passes instructions and changes to orders on to relief guard.	_____	_____

References:
Required: TC 3-22.6,

052-COM-1361

Camouflage Yourself and Your Individual Equipment

Conditions: Given an individual weapon, grass, bushes, and trees, pieces of the Lightweight Camouflage Screen System (LCSS), skin paint, and charcoal and/or mud. Some iterations of this task should be performed in MOPP 4.

Standards: Camouflage yourself and your individual equipment to prevent detection by visual, near-infrared, infrared, ultraviolet, radar, acoustic, and radio sensors.

Special Condition: None

Safety Risk: Low

Special Standards: None

Special Equipment:

MOPP 4: Sometimes

Task Statements

Cue: None

Note:

Performance Steps

1. Apply camouflage principles throughout camouflaged operations.
   a. Employ realistic camouflage.
   　(1) Employ camouflage material that resembles the background.
   　(2) Employ camouflage subtly without overdoing.
   b. Apply camouflaged movement technique.

Note: Movement draws attention, and darkness does not prevent observation. The naked eye and infrared/radar sensors can detect movement.

    (1) Minimize movement.

    (2) Move slowly and smoothly when movement is necessary.

   c. Breakup regular shapes.

    (1) Use natural or artificial materials to breakup shapes, outlines, and equipment.

    (2) Stay in shadows when moving, if possible.

    (3) Disguise or distort the shape of your helmet and your body with natural or artificial materials when conducting operations close to the enemy.

   d. Reduce possible shine by covering or removing items that may reflect light.

Note: Examples of items that should be covered and/or removed include: mirrors, eye glasses, watch crystals, plastic map cases, starched uniforms, clear-plastic garbage bags, red-filtered flashlights, goggles worn on top of helmets cigarettes and pipes.

   e. Blend colors with the surroundings or, at a minimum, ensure that objects do not contrast with the background (figure 052-COM-1361-1).

Note: Change camouflage, as required, when moving from one area to another. What works well in one location may draw fire in another

| Sand and light green for desert and dry areas | Loam and white for snow covered areas | Loam and light green for vegetated areas |

Figure 052-COM-1361-1.
Colors Used for Camouflage

   f. Employ noise discipline.

  2. Camouflage your exposed skin.

Note: Exposed skin reflects light.

   a. Cover your skin oils, using paint sticks, even if you have very dark skin.

Note: Paint sticks cover these oils and provide blending with the background.

   b. Use the color chart in table 052-COM-1361-1 when applying paint on the face.

Table 052-COM-1361-1. Color Chart.

Camouflage Material	Skin Color	Shine Areas	Shadow Areas
	Light or Dark	Forehead, Cheekbones, Ears, Nose, and Chin	Around Eyes, Under Nose, and Under Chin
Loam and Light Green Stick	All troops use in areas with green vegetation	Use loam	Use light green
Sand and Light Green Stick	All troops use in areas lacking green vegetation	Use light green	Use sand
Loam and White Stick	All troops use only in snow covered terrain	Use loam	Use white
Burnt Cork, Bark Charcoal, or Lamp Black	All troops use if camouflage sticks are not available	Use	Do not use
Light – Color Mud	All troops use if camouflage sticks are not available	Do not use	Use

c. Paint high, shiny areas (forehead, cheekbones, nose, ears, and chin) with a dark color

d. Paint low, shadow areas (around the eyes, under the nose and under the chin) with a light color.

CAUTION

Mud contains bacteria, some of which is harmful and may cause disease or infection. Mud should be considered as a last resort for field expedient paint.

Expedient paint containing motor oil should be used with extreme caution. Prolonged exposure to motor oil may result in personal injury.

---

e. Paint exposed skin on the back of the neck, arms, and hands with an irregular pattern.

3. Camouflage your uniform and helmet.

a. Roll your sleeves down, and button all buttons

CAUTION

Soldiers must be aware of local foliage hazards, and possible reactions to poisonous leaves.

---

b. Attach leaves, grass, small branches, or pieces of LCSS to your uniform and helmet (figure 052-COM-1361-2). These items will distort shapes and blend colors with the natural background

Note: ACUs provide visual and near-infrared camouflage

**Figure 052-COM-1361-2.
Camouflaged Helmets.**

c. Wear unstarched ACUs.

Note: Starch counters the infrared properties of the dyes.

d. Replace excessively faded and worn ACUs because camouflage effectiveness is lost.

4. Camouflage your personal equipment

a. Cover or remove shiny items.

b. Secure items that rattle or make noise when moved or worn.

c. Breakup the shape of large and bulky equipment using natural items and/or LCSS.

5. Maintain camouflage.

a. Replace natural camouflage as it dies and loses its effectiveness.

b. Replace camouflage as it fades.

c. Replace camouflage to correspond to changing surroundings.

Evaluation Preparation:

Setup: Provide the Soldier with the equipment and or materials described in the conditions statement.

Brief Soldier: Tell the Soldier what is expected of him by reviewing the task standards. Stress to the Soldier the importance of observing all cautions, warnings, and dangers to avoid injury to personnel and, if applicable, damage to equipment.

Performance Measures	GO	NO GO
1 Applied camouflage principles throughout camouflaged operations.	＿＿	＿＿
a. Employed realistic camouflage.	＿＿	＿＿
b. Applied camouflaged movement technique.	＿＿	＿＿
c. Broke-up regular shapes.	＿＿	＿＿
d. Reduced possible shine by covering or removing items that may reflect light.	＿＿	＿＿
e. Blended colors with the surroundings or, at a minimum, ensured that colors so not contrast with the background.	＿＿	＿＿
f. Employed noise discipline.	＿＿	＿＿
2 Camouflaged your exposed skin.	＿＿	＿＿
a. Covered your skin oils, using paint sticks, even if you have very dark skin.	＿＿	＿＿
b. Used the color chart in table 052-COM-1361-1 when applying paint on the face.	＿＿	＿＿
c. Painted high, shiny areas (forehead, cheekbones, nose, ears, and chin) with a dark color.	＿＿	＿＿

Performance Measures	GO	NO GO
d. Painted low, shadow areas (around the eyes, under the nose and under the chin) with a light color.	___	___
e. Painted exposed skin on the back of the neck, arms, and hands with an irregular pattern.	___	___
3 Protected yourself against physical and other hazards.	___	___
a. Rolled your sleeves down, and buttoned all buttons.	___	___
b. Attached leaves, grass, small branches, or pieces of LCSS to your uniform and helmet.	___	___
c. Wore unstarched ACUs.	___	___
d. Replaced excessively faded and worn ACUs because camouflage effectiveness is lost.	___	___
4 Camouflaged your personal equipment.	___	___
a. Covered or removed shiny items.	___	___
b. Secured items that rattle or make noise when moved or worn.	___	___
c. Broke-up the shape of large and bulky equipment using natural items and/or LCSS.	___	___
5 Maintained camouflage.	___	___
a. Replaced natural camouflage as it dies and loses its effectiveness.	___	___
b. Replaced camouflage as it fades.	___	___
c. Replaced camouflage to correspond to changing surroundings.	___	___

Evaluation Guidance: Score the Soldier GO if all performance measures are passed. Score the Soldier NO-GO if any performance measure is failed. If the Soldier scores a NO-GO, show the Soldier what was done wrong and how to do it correctly.

References
Required: ATP 3-37.34, TC 3-21.75
Related:

Construct Individual Fighting Positions

Conditions: You are a member of a squad that has just occupied a defense position and you have been directed to construct an individual fighting position. You have your assigned weapon(s) (M249 machine gun, M240B machine gun, M16-series rifle, M4- series carbine, and/or a shoulder launched missile), a blank DA Form 5517 Standard Range Card, personal protective equipment, construction material, and camouflage material. You have been given your sectors of fire. Some iterations of this task should be performed in MOPP 4.

Standards: Construct a fight position based on leadership direction and type of weapon(s) assigned. Ensure fighting position provides frontal, side, rear, and overhead cover (OHC), as required. Prepare a range card for the position.

Special Condition: None

Safety Risk: Low

MOPP 4: Sometimes

Task Statements

Cue: None

Note: A fighting position provides cover from fire and concealment from observation while allowing you to engage the enemy with your weapon. There are two types of fighting position: hasty and deliberate. The type of fighting position you construct is dependent on: time available, equipment available, and the required level of protection required. If assigned an M4 rather than an M16-series weapon, add 7 inches (18 centimeters). The length of two M16s is equal to two and a half M4s. The widths of all the fighting positions are only an approximate distance and based on the individual Soldier's equipment.

OHC can be built up or down, this task covers built up OHC. Built-up OHC is constructed on top of the parapets up to 18 inches (46 centimeters) and provides for maximum room inside the fighting position and adequate space between the end walls of the fighting position and the OHC. Built-down OHC is constructed at or below ground level and should not exceed 12 inches (30 centimeters) above ground. This lowers the profile of the fighting position, which aids in avoiding detection. However, it restricts the fighting space between the end walls of the fighting position and the OHC. To account for this restricted space the width of the fighting position should be extended to three M16 lengths.

1. Construct a hasty fighting position

Note: A hasty fighting position should give frontal cover from enemy direct fire but allow firing to the front and the oblique. Hasty positions are used if: there is little time for preparation, there is no requirement for a deliberate defensive position (such as a pause during movement) or you have just occupied the position. A hasty fighting position uses whatever cover is available. The position can be developed into a deliberate position, if in a suitable location.

    a. Construct a shell crater.

        (1). Lie prone in the depression.

        (2). Orient your position so you are oblique to enemy fire.

    b. Construct a skirmisher's trench.

        (1). Physical with firearms used.

Note: A skirmisher's trench is used for immediate shelter from enemy fire when there are no defilade firing positions available. In all but the hardest ground, you can use this technique to quickly form a shallow, body-length pit

        (1) Lie prone or on your side.

        (2). Report the situation immediately to the section or team leader.

        (3). Scrape the soil underneath or beside you with an entrenching tool.

        (4). Pile the soil in a low parapet between yourself and the enemy

    c. Construct a prone fighting position (Figure 071-COM-4408-1).

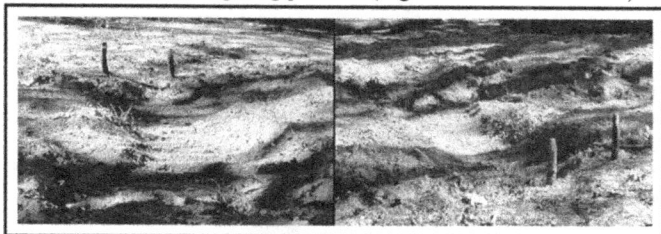

Figure 071-COM-4408-1.
Example of a prone fighting position (Hasty)

        (1). Construct a crater or skirmisher's trench fighting position.

        (2). Scrape additional soil from your position to a depth of about 18 inches (46 centimeters).

        (3). Build cover around the edge of the position by using the dirt dug from the hole.

2. Construct a deliberate fighting position.

    a. Construct a one-man fighting position.

Note: Except for its size, a one-man position is built the same way as a two-man fighting position. The hole of a one-man position is only large enough for you and your equipment. It does not have the security of a two-person position; therefore, it must allow you to shoot to the front or oblique from behind frontal cover.

    b. Construct a two-man fighting position. (Figure 071-COM-4408-2).

Note: A two-man fighting position is preferred over the one-man fighting position as it allows more flexibility and better security. A two-man fighting position is constructed in four stages with the chain of command normally inspecting and providing additional guidance between each phase.

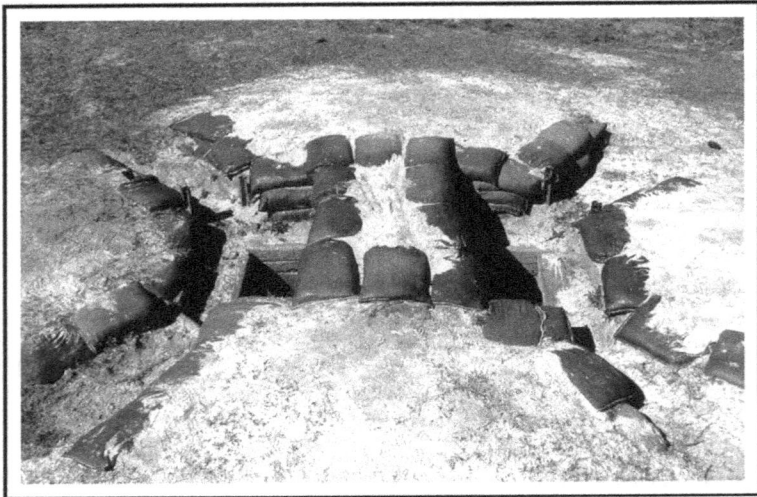

Figure 071-COM-4408-2.
Two-man fighting position with OHC.

(1). Construct stage 1 of a two-man fighting position. (Figure 071-COM-4408-3).

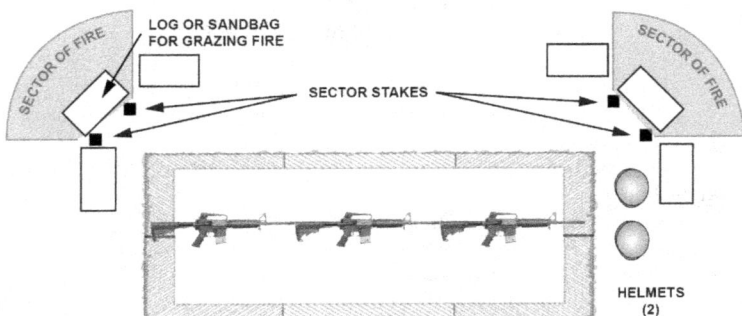

Figure 071-COM-4408-3.
Two-man fighting position - stage 1 (top view).

(a). Identify sector(s) of fire (at least primary and possibly secondary).

(b). Check fields of fire from the prone position.

(c). Emplace sector stakes (right and left) to define your sectors of fire.

Note: The sector stakes must be sturdy and stick out of the ground at least 18 inches (46 centimeters); this will prevent your weapon from being pointed out of your sector.

(d). Emplace aiming and limiting stakes as needed.

Note: Aiming and limiting stakes help you fire into dangerous approaches at night and at other times when visibility is poor. Forked tree limbs about 12 inches (30 centimeters) long make good stakes. One stake (possibly sandbags) is placed near the edge of the hole to rest the stock of your rifle on. The other stake is placed forward of the rear (first) stake/sandbag toward each dangerous approach. The forward stakes are used to hold the rifle barrel.

(e). Emplace grazing fire logs or sandbags to achieve grazing fire 1 meter above ground level.

(f). Scoop out elbow holes to keep your elbows from moving around when you fire.

(g). Trace position outline.

Note: The length of two M16s is equal to two and a half M4s. The widths of all the fighting positions are only an approximate distance and based on the individual Soldier's equipment.

(h). Clear primary and secondary (if applicable) fields of fire

(2). Construct stage 2 of a two-man fighting position. (Figure 071-COM-4408-4).

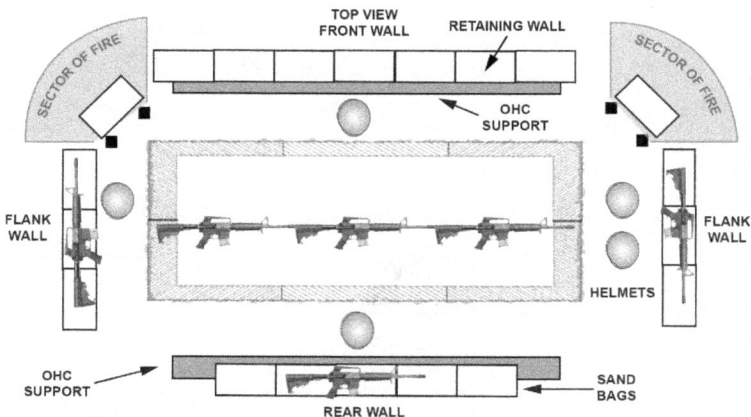

Figure 071-COM-4408-4
Two-man fighting position - stage 2 (top view).

   (a). Emplace OHC supports to front and rear of position, at least 12 inches (30 centimeters) from the edge of the position outline.
Note: 12 inches (30 centimeters) is about 1-helmet length
If you plan to use logs or cut timber, secure them in place with strong stakes from 2 to 3 inches (5 to 7 centimeters) in diameter and 18 inches (46 centimeters) long. Short U-shaped pickets will work.
   (b). Construct parapet retaining walls.
    1. Construct the front retaining wall at least 10 inches (25 centimeters) high, two filled sandbags deep, and equal length of the fighting position.
    2. Construct rear retaining wall--At least 10 inches (25 centimeters) high, and one M16 long.
    3 Construct flank retaining walls--At least 10 inches (25 centimeters) high, and equal width of the fighting position.
   (c). Remove the top layer of dirt from the hole.
    1. Set aside grass or foliage for camouflage
    2. Use excavated soil to fill sandbags
   (3). Construct stage 3 of a two-man fighting position
   (a). Dig position with vertical walls to a maximum depth of armpit deep (if soil conditions permit). (Figure 071-COM-4408-5)
Note: If the walls of the position are unstable, due to soil properties, you can use revetments and/or slope the walls. Plywood or sheeting material and pickets can be used to revet walls. For sloped walls you would first dig a vertical hole and then slope the walls at 1:4 ratio (move 12 inches [30 centimeters] horizontally for each 4 feet [1.22 meters] vertically).

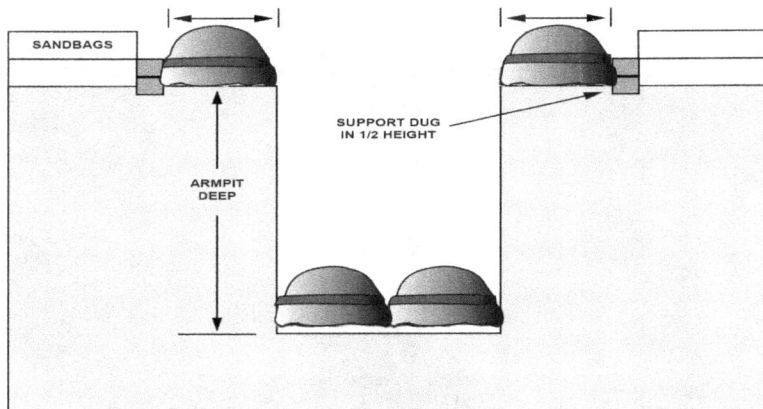

Figure 071-COM-4408-5. Digging the position (side view)

   (b). Use excavated soil from hole to fill parapets in the order of front, flanks, and rear.

(c). Verify you can cover the entire sector of fire from this position

(d). Dig two grenade sumps in the floor one on each end.

Note: Grenade sumps are as wide as the entrenching tool blade; at least as deep as an entrenching tool and as long as the position floor is wide.

(e). Slope the floor toward the grenade sumps.

(f). Dig a storage compartment in the bottom of the back wall; the size of the compartment depends on the amount of equipment and ammunition to be stored.

(g). Install revetments, if required, to prevent wall collapse/cave-in.

(h). Emplace standard length stringers for OHC (Figure 071-COM-4408-6).

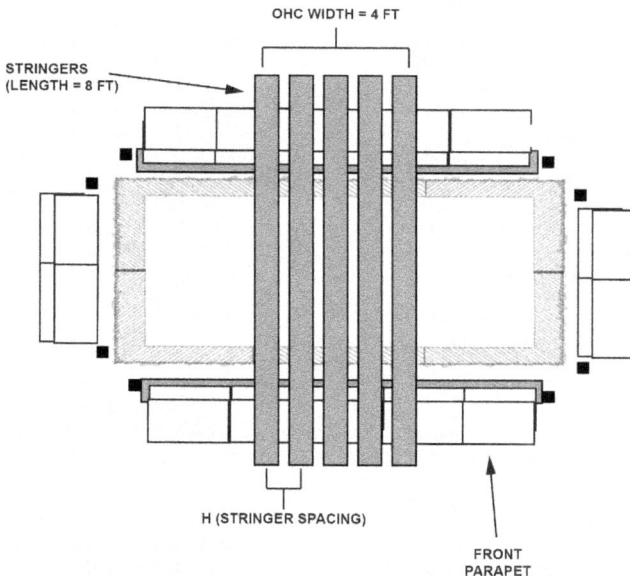

Figure 071-COM-4408-6
Placement of stringers for OHC

(4). Construct stage 4 of a two-man fighting position.

(a). Install OHC. (Figure 071-COM-4408-7).

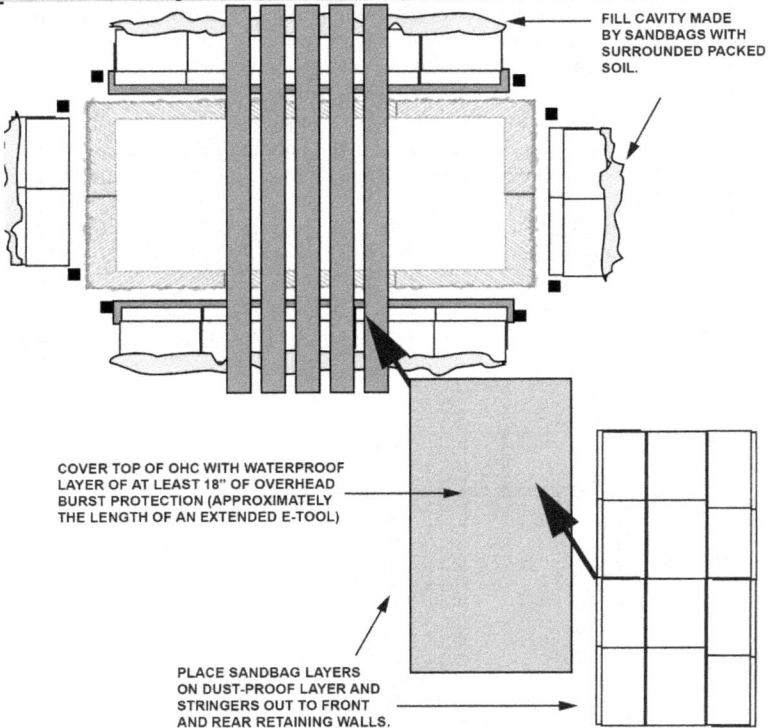

FILL CAVITY MADE
BY SANDBAGS WITH
SURROUNDED PACKED
SOIL.

COVER TOP OF OHC WITH WATERPROOF
LAYER OF AT LEAST 18" OF OVERHEAD
BURST PROTECTION (APPROXIMATELY
THE LENGTH OF AN EXTENDED E-TOOL)

PLACE SANDBAG LAYERS
ON DUST-PROOF LAYER AND
STRINGERS OUT TO FRONT
AND REAR RETAINING WALLS.

Figure 071-COM-4408-7
Installation of OHC

       1. Emplace dustproof layer.
Note: Plywood, sheeting mats can be used as a dustproof layer (could be boxes, plastic panel, or interlocked U-shaped pickets). A standard dustproof layer is 4'x4' sheets of ¾-inch plywood centered over dug position
       2. Nail plywood dustproof layer to stringers, if required
       3. Emplace at least 18 inches (46 centimeters) of filled sandbags for overhead burst protection (***Note:*** At a minimum four layers.) the sandbags must cover the area between the front and rear retaining wall.
       4. Use plastic or a poncho for waterproofing layer.
       5. Fill center cavity with soil from dug hold and surrounding soil.
     (b). Camouflage the fighting position.
       1. Mold the OHC and parapets to blend with the surrounding terrain
       2. Camouflage the position with natural materials that do not have to be replaced

Note: Rocks, logs, live bushes, grass, and other available materials can be used to make the position blend with surroundings, or camouflage screen systems

        3. Ensure the position cannot be seen within 115 feet (35 meters).

    3. Construct a machine gun fighting position. (Figure 071-COM-4408-8 and Figure 071-COM-4408-9)

071-COM-4408-8.
Machine gun fighting position with OHC.

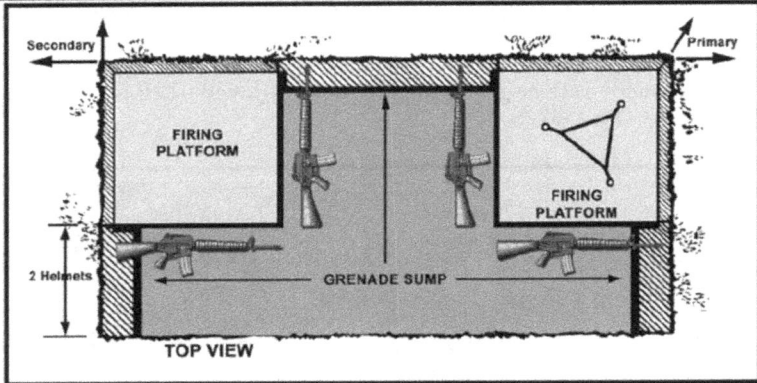

071-COM-4408-9.
Machine gun fighting position (top view).

a. Construct stage 1 of a machine gun fighting position.

(1). Establish sectors (primary and secondary) of fire

(a) Check fields of fire from the prone position.

(b) Assign sector of fire (primary and secondary) and final protective line (FPL) or principal direction of fire (PDF).

(c) Emplace aiming stakes.

(d) Decide whether to build OHC up or down, based on potential enemy observation of position.

(2). Mark the outline of the position.

(a) Trace position outline to include location of two distinct firing platforms.

(b) Mark position of the tripod legs where the gun can be laid on the FPL or PDF.

(3). Clear primary and secondary fields of fire.

b. Construct Stage 2 of a machine gun fighting position.

(1). Dig firing platforms 6 to 8 inches (15 to 20 centimeters) deep and one M16 in length and width.

(2). Emplace the OHC supports to front and rear of the position.

Note: The supports are placed the same as for a two-man fighting position.

(3). Construct the parapet retaining walls.

Note: The parapet retaining walls are constructed the same as for a two-man fighting position.

(4). Position the machine gun to cover primary sector of fire.

c. Construct stage 3 of a machine gun fighting position.

(1) Dig position and build parapets.

(a) Dig the position to a maximum armpit depth around the firing platform.

(b) Use soil from hole to fill parapets in order of front, flanks, and rear.

(c) Dig grenade sumps and slope floor toward them.

(d) Install revetment if needed.

Note: Follow same steps as for two-man fighting position.

(2) Place stringers for OHC.

Note: Stringers are placed the same way as for a two-man position.

d. Construct stage 4 of a machine gun fighting position.

(1) Install OHC.

Note: Build the OHC the same as you would for a two-man fighting position.

(2) Install camouflage.

(a) Use surrounding topsoil and camouflage screen systems.

(b) Ensure position cannot be seen within 115 feet (35 meters).

(c) Use soil from hole to fill sandbags and OHC cavity, or to spread around and blend position in with surrounding ground.

4. Construct a shoulder launched missile fighting position.

a. Construct an M136 fighting position

Note: An M136 fighting position is a standard two-man fighting position that includes basic considerations for firing shoulder launched missile. The shoulder launched missile is fired from a modified standing position by leaning against the rear wall of the fighting position and ensuring the rear of the weapon extends beyond the rear of the fighting position.

(1) Construct stage 1.

Note: Only additional consideration is identifying the backblast area to ensure it is kept cleared. Leaders must ensure that shoulder launched missiles are positioned so that the backblast misses other fighting positions.

(2) Construct stage 2.

Note: Only additional consideration is the rear parapet does not block the backblast area.

(3) Construct stage 3.

Note: No additional considerations.

(4) Construct stage 4.

Note: Only additional consideration is ensuring any camouflage in the backblast area is secure and not easily combustible.

b. Construct a standard Javelin fighting position with OHC.

Note: The standard Javelin fighting position has cover to protect you from direct and indirect fires. The position is prepared the same as the two-man fighting position with two additional steps. See Figure 071-COM-4408-10.

ENEMY FORCES

FRONT PARAPET

JAVELIN PARAPET

GRENADE SUMP

TWO MAN FIGHTING POSITION (STANDING PORTION)

JAVELIN PARAPET

SIDE PARAPET

PRIMARY SEATED FIRING PLATFORM

SECONDARY SEATED FIRING PLATFORM

SIDE PARAPET

STORAGE AREA

LEGEND:

Javelin seated firing platform (1-1/2 M16s wide by 1 M16 long).

Two-man fighting position (2 M16s wide by 1-1/2 M16s long).

Parapet (minimum of 46 centimeters (18 inches) between you and the enemy at least 25 centimeters (10-inches) high and two M16s long).

Overhead cover (1-1/2 M16s wide by 1-1/2 to 2 M16s long) will support at least 46 centimeters (18 inches) of dirt.

Storage area (depends on amount of equipment to be stored, to include extra Javelin rounds).

Grenade sump (1 entrenching tool in diameter and the length of the entrenching tool deep).

Backblast berm is about 1 Kevlar helmet thick and 46 centimeters (18 inches) high. The berm deflects the hot gases and debris up and out. This reduces the amount of clearing required.

Javelin parapet, sand bags if available, should be used.

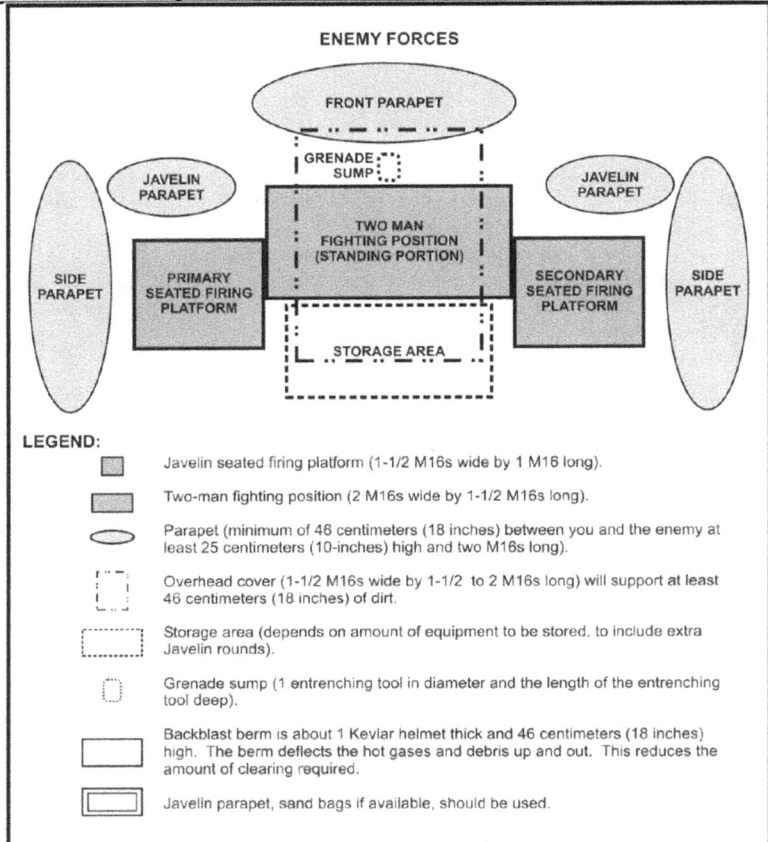

Figure 071-COM-4408-10
Standard Javelin firing position

(1) Extend and slope the back wall of the position rearward to serve as a storage area.

(2) Extend the front and side parapets twice the length as the dimensions of the two-man fighting position with the Javelin's primary and secondary seated firing platforms added to both sides.

5. Prepare a DA Form 5517, Standard Range Card for the fighting position.

Note: A range card is comprised of sectors of fire, principal direction of fire, final protective live, and dead space.

a. Orient the card so both the primary and secondary sectors of fire can fit on it.

b. Draw a rough sketch of the terrain to the front of your position.

Note: Include any prominent natural and man-made features that could be likely targets.

    c. Draw your position at the bottom of the sketch.

Note: Do not put in the weapon symbol at this time.

    d. Fill in the marginal data.

        (1) Gun number or squad.

        (2) Platoon, company and date.

        (3) Magnetic north arrow.

    e. Sketch in the magnetic north arrow on the card with its base starting at the top of the marginal data section.

    f. Using your compass, determine the azimuth in degrees from the terrain feature to the gun position.

    g. Determine the distance between the gun and the feature by pacing or plotting the distance on a map.

    h. Sketch in the terrain feature on the card in the lower left or right hand corner.

    i. Connect the sketch of the position and the terrain feature with a barbed line from the feature to the gun.

    j. Write in the distance in meters.

    k. Add final protective fires to your range card.

        (1) Sketch in the limits of the primary sector of fire as assigned by your leader.

        (2) Sketch in the FPL line on your sector limit as assigned.

        (3) Determine dead space on the final protective line by having your assistant gunner walk the final protective line.

        (4) Watch him walk down the line and mark spaces that cannot be grazed.

        (5) Sketch dead space by showing a break in the symbol for an FPL, and write in the range to the beginning and end of the dead space.

        (6) Label all targets in your primary sector in order of priority.

    l. Prepare range card when assigned a PDF instead of an FPL.

        (1) Sketch in the limits of the primary sector of fire as assigned by your leader.

Note: Sector should not exceed 875 mils, the maximum traverse of the tripod-mounted machine gun.

        (2) Sketch in the symbol for an automatic weapon oriented on the most dangerous target within your sector *Note:* The PDF will be target number one in your sector. All other targets will be numbered in priority.

        (3) Sketch in your secondary sector of fire.

Note: The secondary sector is drawn using a broken line.

        (4) Label targets within the secondary sector with the range in meters from your gun to each target.

Note: When necessary the bipod is used to engage targets in your secondary sector.

        (5) Sketch in aiming stakes, if used.

Evaluation Preparation:

Setup: Provide the Soldier with the equipment and or materials described in the conditions statement.

Brief Soldier: Tell the Soldier what is expected of him by reviewing the task standards. Stress to the Soldier the importance of observing all cautions, warnings, and dangers to avoid injury to personnel and, if applicable, damage to equipment.

Performance Measures	**GO**	**NO GO**
1 Constructed a hasty fighting position.	____	____
2 Constructed a deliberate fighting position.	____	____
3 Constructed a machine gun fighting position.	____	____
4 Constructed a shoulder launched missile fighting position.	____	____
5 Prepared a DA Form 5517, Standard Range Card for the fighting position.	____	____

Evaluation Guidance: Score the Soldier GO if all performance measures are passed. Score the Soldier NO-GO if any performance measure is failed. If the Soldier scores a NO-GO, show the Soldier what was done wrong and how to do it correctly.

References:
Required: DA FORM 5517, TC 3-21.75, TM 3-23.25
Related:

159-COM-2026

Identify Combatant and Non-Combatant Personnel and hybrid Threats

Conditions: In a field, military operations in urban terrain (MOUT), or garrison environment, where a Soldier is required to demonstrate an understanding of the various personnel in an Operational Environment (OE). Standard MOPP 4 conditions do not exist for this task. See the MOPP 4 statement for specific conditions

Standards: Identify the combatant and non-combatant personnel and hybrid threats within an OE.

Special Condition: None

Safety Risk: Low

MopP 4: N/A

Task Statements

Remarks: None

Notes: None

Performance Steps

1. Identify the combatant and/or noncombatant personnel within an OE.
   a. Identify Armed Combatants:
   (1) Regular military forces.
   (2) Internal security forces.
   (3) Insurgent organizations.
   (4) Guerilla organizations.
   (5) Private security organizations.
   (6) Criminal organizations.
   b. Identify Unarmed Combatants
   (1) Unarmed nonmilitary personnel who may decide to support hostilities-recruiting, financing, intelligence-gathering, providing targeting information, supply brokering, transportation, courier, information warfare (videographers), improvised explosive device (IED) fabricators.
   (2) Unarmed combatants may possibly be affiliated with paramilitary organizations.
   (3) Includes support that takes place off the battlefield.

(4) Other examples of unarmed combatants-medical teams, media (local, national, international), non-governmental organizations/private voluntary organizations (NGOs/PVOs), Trans-national corporations, foreign government and diplomatic personnel, internally displaced persons (IDPs), transients, local populace.

    c. Identify the following types of Noncombatants.

        (1) Media personnel.

        (2) Humanitarian Relief Organizations.

        (3) Mult national Corporations.

        (4) Criminal organizations.

        (5) Private Security Organizations.

        (6) Other Noncombatants and Civilian Population Support.

        (7) Information Warfare elements.

  2. Identify Hybrid Threats within an OE

(Asterisks indicates a leader performance step.)

Evaluation Guidance: Score the soldier GO if all performance measure are passed. Score the soldier NO-GO if any performance measure is failed. If the soldier scores NO-GO, show the soldier what was done wrong and how to do it correctly.

Evaluation Preparation: Setup: This task may be evaluated at the end of OE training as well as during a field training exercise.

Brief Soldier: Tell soldier that they will be evaluated on their ability to correctly identify examples of combatant and/or noncombatant personnel and hybrid threats within an OE.

Performance Measures	GO	NO GO
1 Checked the casualty for signs and symptoms of shock.	____	____
2 Positioned casualty correctly.	____	____
3 Loosened clothing at the neck, waist, or anywhere it was binding.	____	____
4 Prevented the casualty from chilling or overheating.	____	____

Performance Measures	GO	NO GO
5 Calmed and reassured the casualty.	____	____
6 Watched the casualty closely for life-threatening conditions and checked for other injuries, if necessary. Sought medical aid.	____	____
7 Sought Medical Aid.	____	____

Environment: Environmental protection is not just the law but the right thing to do. It is a continual process and starts with deliberate planning. Always be alert to ways to protect our environment during training and missions. In doing so, you will contribute to the sustainment of our training resources while protecting people and the environment from harmful effects. Refer to the current Environmental Considerations manual and the current GTA Environmental-related Risk Assessment card. Environmental protection is not just the law but the right thing to do. It is a continual process and starts with deliberate planning. Always be alert to ways to protect our environment during training and missions. In doing so, you will contribute to the sustainment of our training resources while protecting people and the environment from harmful effects. Refer to ATP 3-34.5 Environmental Considerations and GTA 05-08-002 ENVIRONMENTAL-RELATED RISK ASSESSMENT

Safety: In a training environment, leaders must perform a risk assessment in accordance with current Risk Management Doctrine. Leaders will complete the current Deliberate Risk Assessment Worksheet in accordance with the TRADOC Safety Officer during the planning and completion of each task and sub-task by assessing mission, enemy, terrain and weather, troops and support available-time available and civil considerations, (METT-TC). *Note:* During MOPP training, leaders must ensure personnel are monitored for potential heat injury. Local policies and procedures must be followed during times of increased heat category in order to avoid heat related injury. Consider the MOPP work/rest cycles and water replacement guidelines IAW current CBRN doctrine. In a training environment, leaders must perform a risk assessment in accordance with ATP 5-19, Risk Management. Leaders will complete a DD Form 2977 DELIBERATE RISK ASSESSMENT WORKSHEET during the planning and completion of each task and sub- task by assessing mission, enemy, terrain and weather, troops and support available-time available and civil considerations, (METT-TC). *Note:* During MOPP training, leaders must ensure personnel are monitored for potential heat injury. Local policies and procedures must be followed during times of increased heat category in order to avoid heat related injury. Consider the MOPP work/rest cycles and water replacement guidelines IAW TM 3-11.32 Multi-

Service Reference for Chemical, Biological, Radiological, and Nuclear Warning and Reporting and Hazard Prediction Procedures.

References: ADP 3-0; TC 7-100

Skills : None

181-COM-1001

Conduct Operations According to the Law of War

Conditions: Given an overview of the Law of War be able to identify, understand, and comply with the provisions of the Law of War, including the Geneva and Hague Conventions and the 10 Soldier rules. You have access to the operational hand book, FM 27-10, FM 2-22.3, and the Manual for Courts-Martial, and other related materials.

Standards: Identify, understand, and comply with the Law of War. Identify problems or situations that violate the policies and take appropriate action, including notifying appropriate authorities, so that expedient action may be taken to correct the problem or situation.

Special Condition: None

Safety Risk: Low

MOPP 4:

```
                            Task Statements
```

Cue: None

Note: None

Performance Steps

    1. Identify the key elements of the Law of War.
      a. Describe how the Hague Convention and Geneva Conventions pertain to combat operations.
      b. Describe International Customary Law of War.
      c. Describe other international Treaty law.
    2. Describe the responsibilities of U.S. Soldiers to obey the Law of War.
      a. To assist Commanders and Soldiers in mission accomplishment.
      b. To regulate the use of force and prohibit unlawful conduct.

c. To protect against unnecessary suffering and excessive collateral damage.

d. To promote the humane treatment of noncombatants, wounded and sick, and civilians.

3. Identify the basic principles of the Law of War.

a. Define Military Necessity.

(1) Describe a Legitimate Military Target.

(2) Define combatant force.

(3) Decrribe collateral damage.

b. Defines Unnecessary Suffering

c. Define Discrimination and Distinction

d. Define Proportionality.

e. Define Chivalry

4. Identify the "10 Soldier's Rules".

a. Soldiers only fight enemy combatants.

b. Soldiers treat humanely all who surrender or are captured.

c. Soldiers do not kill or torture detained personnel.

(1) List the 5 S's and T.

(2) Describe humane treatment.

(3) Describe respect and protect.

d. Soldiers collect and care for the wounded.

e. Soldiers do not attack protected places or persons.

f. Soldiers do not attack medical personnel, facilities or equipment.

g. Soldiers destroy no more than the mission requires.

h. Soldiers treat civilians and noncombatants humanely.

i. Soldiers do not steal. Soldiers respect private property and possessions.

j. Soldiers should do their best to prevent violations of the Law of War.

k. Soldiers report all violations of the Law of War to their superior.

5. Identify actions to prevent Law of War violations.

a. List actions to protect civilians/noncombatants.

b. List actions to protect civilians/noncombatants.

c. List actions to protect prisoners of war, retained persons and detainees.

d. List actions to protect medical transports and facilities.

e. List actions to prevent engagement of unlawful targets.

f. List actions to prevent excessive use of force.

g. List actions to prevent the unauthorized use of medical service symbols, flag of truce, national emblems, and enemy insignia/uniforms.

h. List actions to prevent unnecessary destruction and seizure of property.

i. List actions to prevent unnecessary suffering and harm.

j. List actions to enforce the rights and responsibilities of EPWs, and detainees.

Evaluation Preparation:

Setup: Evaluate this task at the end of Law of War training.

Brief Soldier: Tell the Soldier that he or she will be evaluated on his or her ability to identify, understand, and comply with the Law of War, including the Geneva and Hague Conventions and the 10 Soldier's Rules. Tell the Soldier that he or she will also be evaluated on his or her ability to identify problems or situations that violate the Law of War and take appropriate action to prevent Law of War violations do not occur.

Performance Measures	GO	NO GO
1 Identified the key elements of the Law of War.	_____	_____
2 Described the responsibilities of U.S. Soldiers to obey the Law of War.	_____	_____
3 Identified the basic principles of the Law of War.	_____	_____
4 Identified the "10 Soldier's Rules".	_____	_____
5 Identified actions to prevent Law of War violations.	_____	_____

Evaluation Guidance: Score the Soldier GO if all performance measures are passed. Score the Soldier NO GO if any performance measure is failed. If the Soldier scores NO GO, show the Soldier what was done wrong and how to do it correctly.

References:
Required:

171-COM-0011

Employ Progressive Levels of Individual Force

WARNING
During the assessment ensure that biological threats associated with close contact/combat are taken into consideration and protective measures are taken to prevent exposure.

Conditions: You are a member of a section or team that is securing a critical area or defusing a civil disturbance and you are approached/confronted by one or more hostile civilians. You have your individual weapon, personal protection equipment (PPE), and the rules of engagement (ROE).

Standards: Assess and immediately report threats situations to your leadership. Protect yourself against hazards. Isolate hostile civilians, if required. Control the situation using the least amount of force possible.

Special Condition: None

Special Standards: None

Special Equipment:

Task Statements

Cue: None

*Note:*The operational environment must be considered at all times during this task. All Army elements must be prepared to enter any environment and perform their missions while simultaneously dealing with a wide range of unexpected threats and other influences. Units must be ready to counter these threats and influences and, at the same time, be prepared to deal with various third-party actors, such as international humanitarian relief agencies, news media, refugees, and civilians on the battlefield. These groups may or may not be hostile to us, but they can potentially affect the unit's ability to accomplish its mission.

Performance Steps
1. Assess the situation by identifying the level of hostile civilian threat.
   a. Verbal.
   b. Physical without weapons (touching, pushing).
   c. Physical with weapons (rocks, clubs, spitting).
   d. Physical with firearms shown.
   e. Physical with firearms used.
2. Report the situation immediately to the section or team leader.
3. Protect yourself against physical and other hazards.
   a. Use full - face shields.
   b. Use double layer latex gloves.
Note: Any exposure incident must be reported to the chain of command.
4. Isolate hostile civilian(s), if required.
   a. Identify hostile group(s) sphere of influence.
   b. Remove the individual with the most influence of the crowd.
   c. Use the 5S's (Search, Silence, Segregate, Safeguard, Speed to the rear).

5. Employ no more force than is necessary to control the situation using graduated response measures.

Note: Soldiers should employ the lowest level of force necessary to address a threat but may use any level, even deadly force, without performing earlier steps, if the circumstances or threat do not allow for the use of graduated levels of force.

    a. Avoid confrontation if possible.

    b. Do not deliberately instigate, threaten, provoke, or bluff.

    c. Speak sternly to the civilian and state the peaceful intent of your mission.

    d. Tell the civilian to "STAND BACK" and warn them that you may have to use force.

    e. If a civilian places his or her hands on your body, brush them back with hand or availble PPE.

    f. If a civilian attempts to inflict bodily harm, use any authorized materials (such as water hoses, chemical gases) to impede movement.

    g. Use your individual weapon, if necessary, as prescribed by the established ROE.

6. Establish and maintain control of the situation.

    a. Comply with the ROE, any host-nation requirements, applicable international treaties and operational agreements.

Note: ROE are directives issued by competent military authority that delineate the circumstances and the limitations under which United States forces will initiate and/or continue combat engagement with other forces encountered. ROE help commanders accomplish the mission by regulating the rules of the use of force. Everyone must understand the ROE and be prepared to execute them properly in every possible confrontation.

    b. Minimize casualties and damage.

Evaluation Preparation:

Setup: Provide the Soldier with the equipment and or materials described in the conditions statement.

Brief Soldier: Tell the Soldier what is expected of him by reviewing the task standards. Stress to the Soldier the importance of observing all cautions, warnings, and dangers to avoid injury to personnel and, if applicable, damage to equipment.

Performance Measures	GO	NO GO
1 Assessed the situation by identifying the level of hostile civilian threat.	___	___

Performance Measures	GO	NO GO
2 Reported the situation immediately to the section or team leader.	_____	_____
3 Protected yourself against physical and other hazards.	_____	_____
4 Isolated hostile civilian(s), as required.	_____	_____
5 Employed no more force than was necessary to control the situation.	_____	_____
6 Established and maintained control of the situation.	_____	_____

Evaluation Guidance: Score the Soldier GO if all performance measures are passed. Score the Soldier NO-GO if any performance measure is failed. If the Soldier scores a NO-GO, show the Soldier what was done wrong and how to do it correctly.

References

Required

Related: ATP 3-22.40, FM 27-10, TC 7-98-1

Related: AR 27-1, FM 27-10

191-COM-0009

Search a Detainee

Conditions: Given your individual equipment, assigned weapon, a detainee, disposable restraints, a guard, an interpreter (if available), DA Forms 4002 (Evidence/Property Tag) and 4137 (Evidence/Property Custody Document), DD Form 2745 (Enemy Prisoner of War (EPW) Capture Tag), and materials to mark and bundle evidence and property. Some iterations of this task should be performed in MOPP 4.

Standards: Search and restrain the detainee sequentially according to the performance steps; locate and confiscate all weapons, contraband, and items of intelligence value; and prepare *DD Form 2745* and *DA Form 4137* without error.

Special Condition: None.

Safety Risk: Low

MOPP 4: Sometimes

Task Statements

Cue: None

WARNING

The searcher must avoid crossing the line of sight or fire of the overwatch during the person search.

CAUTION

Searching a person requires two searchers working together. One searcher conducts the physical search while the other provides overwatch and observes both the searcher and the person being searched. The person providing overwatch should be placed at a 45 degree angle out of the subject's reach. **STAY OUT OF REACH OF THE DETAINEE.**

Performance Steps

1. Position the detainee
   a. Direct the detainee to stand and face you.

Note: If an interpreter is not available, you may have to demonstrate the desired movement to the detainee to overcome the language barrier.

   b. Direct the detainee to raise his/her arms above his/her head, lock his/her elbows, and spread his/her fingers with his/her palms facing you.
   c. Check the detainee's hands visually for weapons or contraband.
   d. Order the detainee to turn around and drop to his/her knees.
   e. Search the back of the detainee's hands for weapons or contraband.
   f. Direct the detainee to lie on his/her stomach, extend his/her arms straight out to the sides with the palms up, and place his/her forehead on the ground.
   g. Tell the detainee to spread his/her legs as far apart as possible, turn his/her feet outward, and keep his/her heels in contact with the ground.
   h. Ensure that the guard remains in front of and at an oblique angle to the detainee (opposite the side being searched).
2. Restrain the detainee.

Note: The situation may also dictate using a blindfold, ear plugs, or a muffle (an item such as cloth to prevent speech or outcry without causing injury) as deemed appropriate or directed by your supervisor.

a. Approach the front of the detainee at about a 45° angle from the side opposite the guard and focus the search on the side of the detainee closer to you

b. Squat and put your knee that is nearer the detainee on the detainee's lower back.

Note: This is done to ensure control, not to inflict pain or discomfort.

c. Direct the detainee to put the arm that is nearer the searcher behind the detainee's back with the palm facing up.

d. Maintain positive control of that arm.

CAUTION

THE DISPOSABLE RESTRAINTS SHOULD BE TIGHT ENOUGH TO SECURE THE DETAINEE'S HANDS, BUT LOOSE ENOUGH TO ALLOW ONE FINGER BETWEEN THE DISPOSABLE RESTRAINTS AND THE DETAINEE'S WRIST TO ENSURE THAT THE DISPOSABLE RESTRAINTS DO NOT RESTRICT THE DETAINEE'S CIRCULATION.

---

e. Grasp the detainee's other hand in a handshake hold, pull it across the top of the hand already under control, apply disposable restraints, and tighten them.

3. Search the detainee.

Note: The body search is the prone frisk search. It is used to quickly detect contraband or weapons that could be used to cause injury or death.

a. Use the bending and crushing technique, remove items as items are discovered, and set them aside.

Note: Conduct same-gender searches when possible. If mixed-gender searches are necessary for speed and security, conduct them in a respectful manner and in the presence of an additional witness to address false claims of misconduct. Further, consider your location at the time of the search and try to use any cover or protective barrier when possible.

(1) Bend the seams to determine if razor blades or similar devices are hidden.

(2) Grasp loose clothing, pull it away from the skin, and squeeze it to detect objects hidden under or within clothing.

(3) Cover each area by repeating the crushing technique until you are sure there are no hidden objects.

b. Announce loudly any weapon found so that the guard and interpreter can clearly hear (for example, gun, knife, or razor).

(1) Alert the guard.

(2) Remain in firm contact with the detainee as you remove the weapon from its hiding place.

(3) Stand up with the weapon, being careful not to walk between the guard and the detainee, and place the weapon a safe distance away within view of the guard.

(4) Return to the detainee and continue searching.

c. Hold the disposable restraints between the detainee's hands and lift his/her arms slightly. Search the area in the small of the back.

d. Release the disposable restraints and stand.

e. Move to the area of the detainee's waist and face the detainee's head, squat (but do not rest your knee on the ground or on the detainee), and pivot (if required) to conduct the rest of the search.

f. Remove the detainee's headgear (if not already removed).

(1) Bend the seams, before crushing, to determine if razor blades or similar devices are hidden.

(2) Place the headgear on the floor or ground.

g. Search the detainee's head and hair.

h. Search the detainee from fingers to shoulders.

(1) Search the collar and neck area (pull dog tags or necklace to the detainee's back).

(2) Remove anything that could be used as a weapon.

i. Search the detainee's back from shoulder to waist on the side nearer the searcher.

(1) Grasp the inside of the detainee's closer elbow.

(2) Pull the detainee upward onto his/her side just high enough to search the front (shoulder to waist) without placing the detainee completely on his/her side.

Note: When searching a female detainee at chest level, the searcher searches down the middle of the bra; around the breast; below the bra; and along the connecting point on the bra and the back, if the clasp is not there, for contraband.

j. Switch hands while controlling the detainee's elbow without changing position.

k. Search the detainee from waist to knee, including the crotch.

Note: Do not be timid while searching the detainee's groin area. Experience has proven that this is a prime location for hiding weapons and contraband. Check it thoroughly.

l. Return the detainee to the facedown position, release the elbow and remind the detainee to keep his/her feet spread and his/her heels on the ground.

m. Direct the detainee to raise his/her leg by bending his/her knee.

n. Grasp the detainee's foot and search from the knee up to the foot. Check the footwear edges and soles.

(1) Check the top of the footwear by inserting a finger in the top edge and feeling for weapons.

(2) Check the footwear edges and soles.

o. Direct the detainee to put the foot back down.

p. Stand and move to the detainee's unsearched side. Move around the detainee's head, but do not walk between the detainee and the guard.

q. Ensure that the guard rotates to the other side of the detainee (the side opposite the side to be searched) while maintaining a 45-degree angle from the detainee's head.

CAUTION

WHEN PREPARING TO TURN DETAINEES OVER, THEY MAY ATTEMPT TO BITE OR SPIT AT THE SEARCHER. ANTICIPATE THIS AND MOVE AS APPROPRIATE TO AVOID SUCH AN ACT.

NOTE: The search is now complete and you have confiscated all material found on the detainee.

---

r. Squat beside the detainee with your body facing the same direction as the detainee's head and search the other side in the same manner as the first.

s. Assist the detainee to stand.

(1) Turn the detainee onto the side facing away from you.

(2) Direct the detainee to bring his/her knees to his/her chest.

(3) Grasp the detainee's arms at the shoulder area and assist him/her to his/her knees.

(4) Pull back on the detainee's arms to help him/her rise to his/her feet.

(5) Ensure that the guard remains focused on the detainee and gathers information as to the detainee's demeanor.

Note: The DD Form 2745 and DA Form 4137 should be completed at the point of capture. However, when you are in imminent danger, these two steps can be completed once you and the detainee are in a safe location.

4. Complete a DD Form 2745.

Note: The DD Form 2745 is a perforated three-part form that is individually serial-numbered. If you run out of DD Forms 2745, use a field expedient method to tag.

a. Ensure that the following minimum information is recorded:

(1) The date and time of capture.

(2) The detainee's name (if known).

Note: Use the DD Form 2745 number as the detainee's name to account for those who are unable or unwilling to provide this information (for example, those who are sick or injured and/or those who do not speak English if an interpreter is not available).

(3) The location of the capture (grid coordinates).

(4) The capturing unit.

(5) The circumstances of the capture, (for example, how the detainee was captured, did the detainee resist, and did the detainee surrender). Record the following minimum information:

Note: Due to the limited space on the DD Form 2745, you may need or be required to document the circumstances of capture on a separate sheet of paper or another form, such as the DA Form 2823 *Sworn Statement.*

Note: Circumstances of capture are essential in determining individual detainee status, making subsequent decisions to release or detain, and collecting and documenting items of intelligence and evidentiary value for custody transfer decisions or future judicial proceedings.

      (a) Various groups, locations, and activities from which the individual detained was operating.

      (b) The physical condition of the detainee.

      (c) The weapons the detainee had, if applicable.

  b. Tag the detainee and his/her equipment. (If you are using a field-expedient method, ensure that you make 3 copies to represent parts A, B, and C of the form.) Distribute the DD Form 2745 tag as follows:

      (1) Ensure that part A is attached to the detainee.

      (2) Retain part B for yourself and/or the unit.

      (3) Ensure that part C is attached to confiscated property (an individual item or attached to a bag or bundle).

  5. Document property/evidence.

    a. Mark all confiscated items with the detainee's DD Form 2745 number using one of the following methods if time permits and materials are available:

      (1) Place the DD Form 2745 number in the "MPR/CIR Sequence Number" block of DA Form 4002 and annotate the item number from the DA Form 4137 and a short description in the "Remarks" block.

Note: this is the preferred method for large items because it does not damage them.

      (2) Place the property/evidence in a resealable bag and mark the outside of the bag with a permanent marker.

Note: This is the preferred method for small items.

      (3) Write the number directly on the property with a permanent marker.

      (4) Etch the number using a sharp object.

Note: Carefully consider how and where to place identification marks on items. Unnecessary damage or destruction of items of personal property or valuable items that may ultimately be returned to the detainee or suspect is unwarranted. To avoid defacing or damaging items, identification markings should be as inconspicuous as possible. Otherwise, place the item in a container that can be sealed and marked.

    b. Prepare DA Form 4137.

      (1) Annotate the DD Form 2745 number in the "MPR/CID Sequence Number" block.

      (2) Insert the name of your unit in the "Receiving Activity" block.

      (3) Place an accurate description of the location where your organization is currently based in the "Location" block (for example, the installation, state, and zip code or the deployed base camp and/or operating base).

(4) Enter the name, grade, and title (if known) of the person who owned or possessed the confiscated items in the block labeled, "Name, Grade, and Title of Person from Whom Received."

(5) Check the "Owner" box if the person or detainee owns the property that you confiscated during the search with the detainee's first name, middle initial, last name, rank, and title.

(6) Check the "Other" box if ownership is unknown. For example, a weapon is discovered by another individual or turned in by another unit at the point of capture.

(7) Enter "N/A" if the property does not come from a specific person (for example, the item is found at a certain location or is collected during a search).

c. Enter the address of the person from whom you received the items in the "Address" block, if known. If it did not come from a person (if it came from a crime scene or point of capture), enter "N/A."

d. Fill in the "Location From Where Obtained" block. If evidence and/or property was obtained from—

(1) A person, enter "person of" and then enter the person's grade and last name and the location where the evidence and/or property was collected. Describe where the item was found on the person (for example, removed from left front pants pocket).

(2) A location, annotate the exact location when the property was found in the area that the detainee was captured (for example, a description might read, "two-story house next to Exxon station on MSR Tampa IVO Baghdad").

e. Enter the reason for confiscation in the "Reason Obtained" block (for example, enter "confiscated during search of the detainee").

f. Record the date-time group of confiscation or item discovery (1400 hrs/15 Sep 06). Indicate the time span when they were collected (1400 hrs-1500 hrs/15 Sep 06) if several items were confiscated. Note the first time when the first item was taken and note the last time when the last item was confiscated.

g. Enter the item numbers. List items consecutively.

h. List the quantity of each item in the "Quantity" column. (Like items may be listed as a group. For example, 20 pills found in a container may be entered as one entry.)

i. Describe each item in the "Description of Articles" block. Describe each item by what can be observed. Use plain bond paper to record the continuation if necessary.

(1) Specify where and how you marked the items for identification (for example, "Marked for ID, 0090829 on barrel").

(2) List the color, size, and shape.

Note: Never list or estimate the value of articles or describe the type of metal or stone in items. For example, describe an item that looks like gold as gold-colored metal.

(3) List serial numbers or identifying marks if available.

(4) Place continuous slashes (////) from the left border of the block to the right border of the block to indicate the end of the list.

j. Complete the "Chain of Custody" portion of the form to transfer items from the owner or individual from whom the item is obtained to the person receiving custody of the items.

(1) Write "1 through 3" in the "Item Number" column if three items are listed in the "Description of Articles" block. The "Chain of Custody" portion of the form is also used to transfer items from one person to another. If only certain items are released, list only those items (for example, "Item 1 and 3").

(2) Enter the date of the custody transfer in the "Date" block.

(3) Fill in the "Released By" column as follows:

(a) Enter the full name in the "Name, Grade, or Title" block if the property is confiscated from an individual. Have the person sign in the "Signature" block. Enter the words "Refused to sign" or "Unable to sign" in the "Signature" block if the person refuses or is unable to sign. There is no legal requirement for the form to be witnessed if the individual refuses to sign.

(b) Enter "N/A" in the "Signature" block if the property does not come from a specific person (for example, if it is obtained from the capture scene).

(4) Fill in the "Received By" column (for example, enter the name, grade, or title of the person taking custody).

(5) Enter the reason for the custody transfer in the "Purpose of Change of Custody" column (for example, "Confiscated from detainee," "Detainee transferred to holding area," or "Detainee transferred to local authorities").

k. Bundle all property (if necessary) and place it in a secure location away from the search area when the search is completed and you have confiscated all the material found on the detainee (allow detainees to keep their helmets, clothing, and any chemical, biological, radiological, and nuclear protective equipment once they have been searched thoroughly).

Note: Any material or method may be used to bundle property, as long as it is secure, will protect the property, and can be marked in such a way that it can be tracked with the detainee's DD Form 2745 number.

Evaluation Preparation:

Setup: Provide the Soldier with role players as a guard and detainee. Provide the detainee role player with props (a knife, handgun, and/or intelligence papers) to guide on his/her person. The guard role player is not absolutely required to evaluate this task but is recommended to add realism.

Brief Soldier: Tell the Soldier to search the detainee according to the performance steps unless otherwise directed by the evaluator. Instruct the Soldier whether you want him/her to fill out the appropriate forms on any items confiscated or to explain to you how he/she would complete them. Tell the guard to provide security and not to assist the searcher in the performance of the task. Tell the detainee to follow the instructions of the soldier and not to resist.

Performance Measures	GO	NO GO
1 Position the Detainee.	_____	_____
2 Restrained the Detainee.	_____	_____
3 Searched the Detainee.	_____	_____
4 Completed a DD Form 2745.	_____	_____
5 Documented property/evidence.	_____	_____

References:

Required: DA FORM 4002; DA FORM 4137; DD FORM 2745; FM 3-63

Related:

071-COM-0512

Perform Hand-to-Hand Combat

Conditions: You are a member of a dismounted squad conducting operations and you have encountered an unarmed adversary. You may be equipped with personnel protective equipment (PPE).

Standards: Dominate the enemy using the basic fighting strategy by achieving the clinch, gaining a dominant position and finishing the fight.

Special Condition: None

Special Standards: None

Safety Risk: Low

Task Statements

Cue: None

Note: This task is only a basic introduction to combatives.

Performance Steps

1. Close the distance.

Note: Controlling a standup fight means controlling the range between fighters. The untrained fighter is primarily dangerous at punching range. The goal is to avoid that range. Even if you are the superior striker, the most dangerous thing you can do is to spend time at the range where the enemy has the highest probability of victory.

    a. Achieve the clinch.

        (1) Face your opponent, and assume the Fighting Stance just outside of kicking range.

        (2) Tuck your chin, and use your arms to cover your head while aggressively closing the distance.

        (3) Drive your head into your opponent's chest.

        (4) Move your cupped hands to your opponent's biceps.

        (5) Aggressively fight for one of the dominant clinch positions.

    b. Achieve the Modified Seatbelt Clinch. (Figure 071-COM-0512-1)

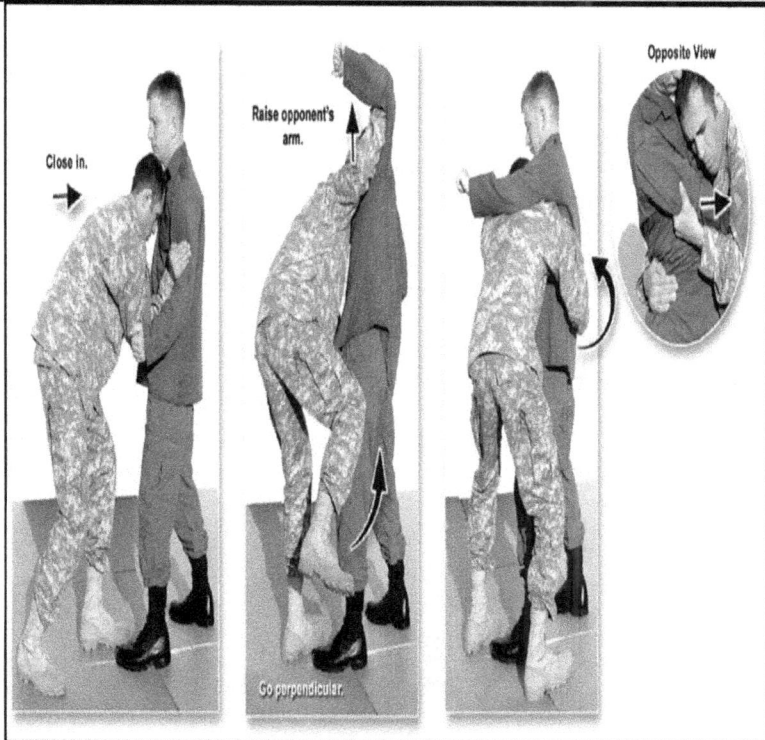

Figure 071-COM-0512-1. Modified Seatbelt Clinch.

    (1) Raise one of his arms.
    (2) Move yourself perpendicular to your opponent.
    (3) Reach around your opponent's waist to grab his opposite-side hip.
    (4) Pull his arm into your chest with your other arm.
    (5) Control his arm at the triceps.
  c. Achieve the Double Under-hooks Clinch. (Figure 071-COM-0512-2)

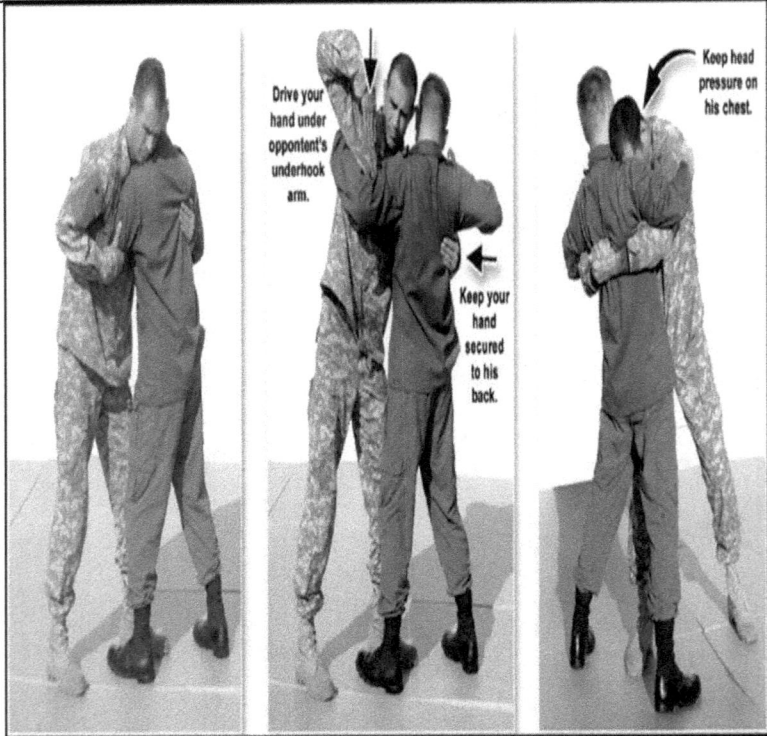

Figure 071-COM-0512-2. Double Underhooks Clinch.

(1) Drive your overhook hand (with a knife edge) under your opponent's underhook arm.

(2) Clasp your hands in a Wrestler's Grip behind your opponent, while keeping head pressure on his chest.

d. Achieve the Rear Clinch. (Figure 071-COM-0512-3)

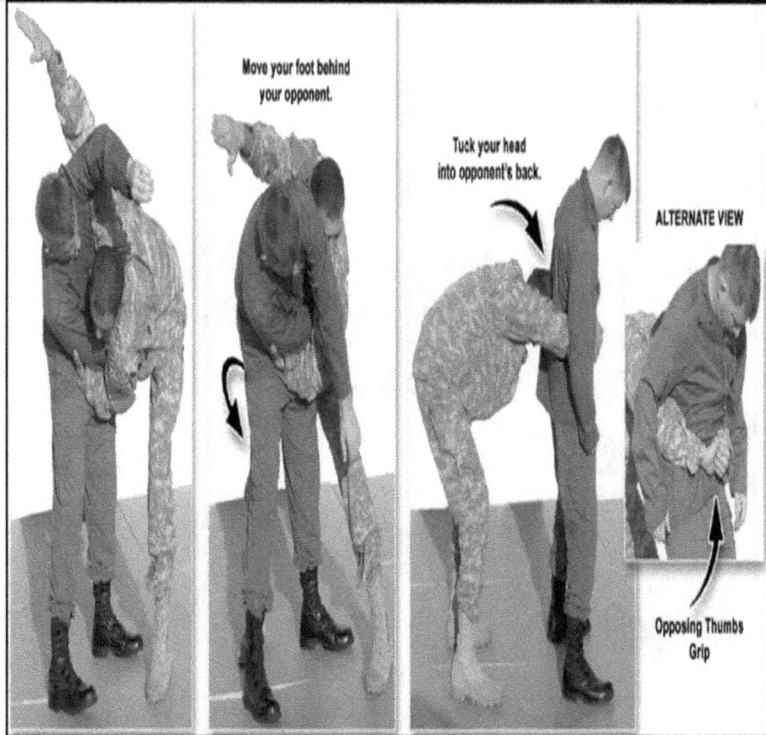

Figure 071-COM-0512-3. The Rear Clinch.

(1) Step behind your opponent.

(2) Clasp your hands around your opponent's waist in an Opposing Thumbs Grip.

(3) Place your forehead in the small of his back to avoid strikes.

Note: From this secure position, you can attempt to take the opponent down.

2. Gain dominant position.

a. Achieve the Rear Mount. (Figure 071-COM-0512-4)

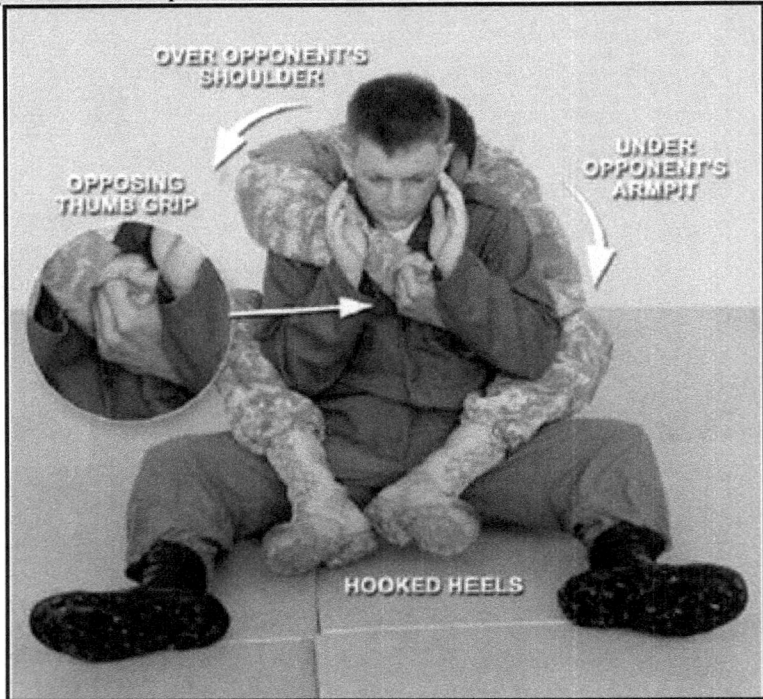

Figure 071-COM-0512-4. The Rear Mount

(1) Place one arm under your opponent's armpit and the other over his opposite shoulder.

(2) Clasp your hands in an Opposing Thumbs Grip.

WARNING

When in the Rear Mount, DO NOT cross your feet; this would provide the opponent an opportunity for an ankle break.

---

(3) Wrap both legs around your opponent, with your heels hooked inside his legs

Note: Keep your head tucked to avoid headbutts.

b. Achieve the Mount. (Figure 071-COM-0512-5)

Note: The Mount allows the fighter to strike the opponent with punches, while restricting the opponent's ability to deliver effective return punches. The Mount also provides the leverage to attack the opponent's upper body with chokes and joint attacks.

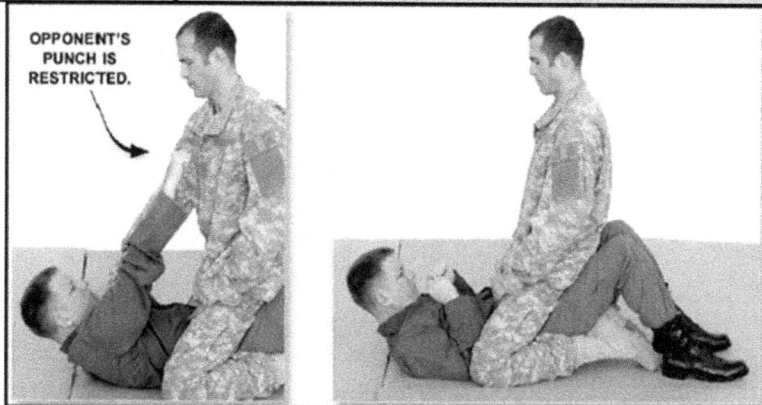

Figure 071-COM-0512-5. The Mount.

(1) Position your knees as high as possible toward the opponent's armpits.

(2) Place your toes in line with or inside of your ankles to avoid injuring your ankles when your opponent attempts to roll you over.

c. Achieve the Guard. (Figure 071-COM-0512-6)

Note: A fighter never wants to be under his opponent; the Guard enables him to defend himself and transition off of his back into a more advantageous position. The Guard allows the bottom fighter to exercise a certain amount of control over the range by pushing out or pulling in his opponent with his legs and hips. With skill, the bottom fighter can defend against strikes and even apply joint locks and chokes.

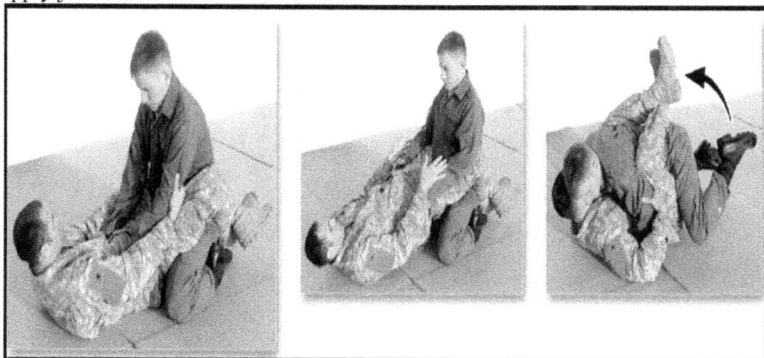

Figure 071-COM-0512-6. The Guard.

(1) Control opponent's arms at the elbows.

(2) Lock your ankles around opponent's torso.

d. Achieve Side Control. (Figure 071-COM-0512-7)

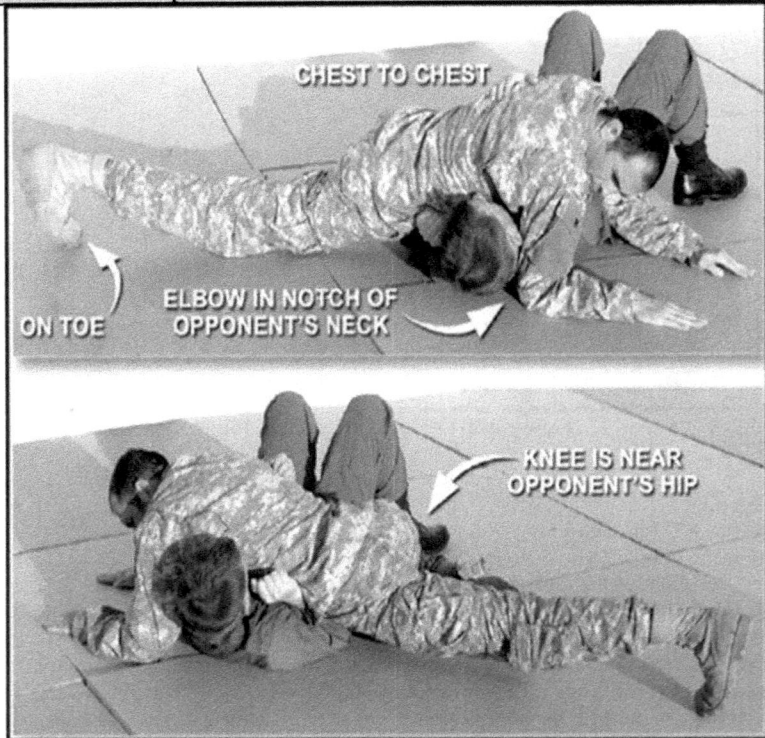

Figure 071-COM-0512-7. Side Control.

(1) Keep the leg closest to your opponent's head straight.

(2) Bend the other leg so that the knee is near your opponent's hip.

(3) Keep your head turned away to avoid knee strikes.

(4) Place your elbow on the ground in the notch created by the opponent's head and shoulder.

(5) Position your other hand palm down on the ground under the opponent's near-side hip.

3. Finish the Fight.

Note: When dominant body position has been achieved, the fighter can begin attempts to finish the fight secure in the knowledge that if an attempt fails, as long as he maintains dominant position, he may simply try again.

a. Achieve the Rear Naked Choke.(Figure 071-COM-0512-8)

Note: The Rear Naked Choke slows the flow of blood in the carotid arteries, which can eventually cause your opponent to be rendered unconscious for a short period of time.

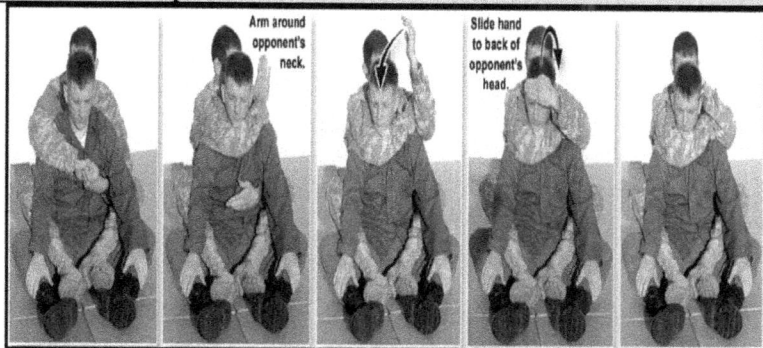

Figure 071-COM-0512-8. Rear Naked Choke.

(1) Place your bicep along one side of your opponent's neck; take your forearm and roll it around to the other side of neck, his chin will line up with your elbow.

(2) Tighten choke up and grab your shoulder or arm.

(3) Place your opposite hand behind the head as if your combing his hair back.

(4) Tuck your head in to avoid getting hit.

(5) Roll your shoulders back, push chest forward and finish the choke.

   b. Achieve the Cross Collar Choke from the Mount and Guard. (Figure 071-COM-0512-9 and 10)

Note: The Cross-Collar Choke is a blood choke that can only be employed when your opponent is wearing a durable shirt. This choke should be performed from either the Mount or Guard.

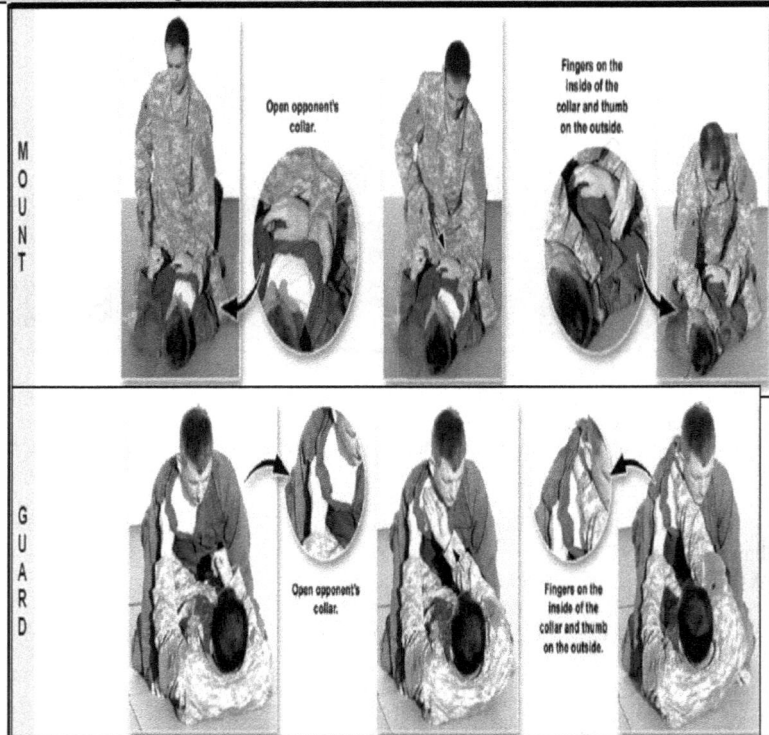

Figure 071-COM-0512-9. Cross Collar Choke.

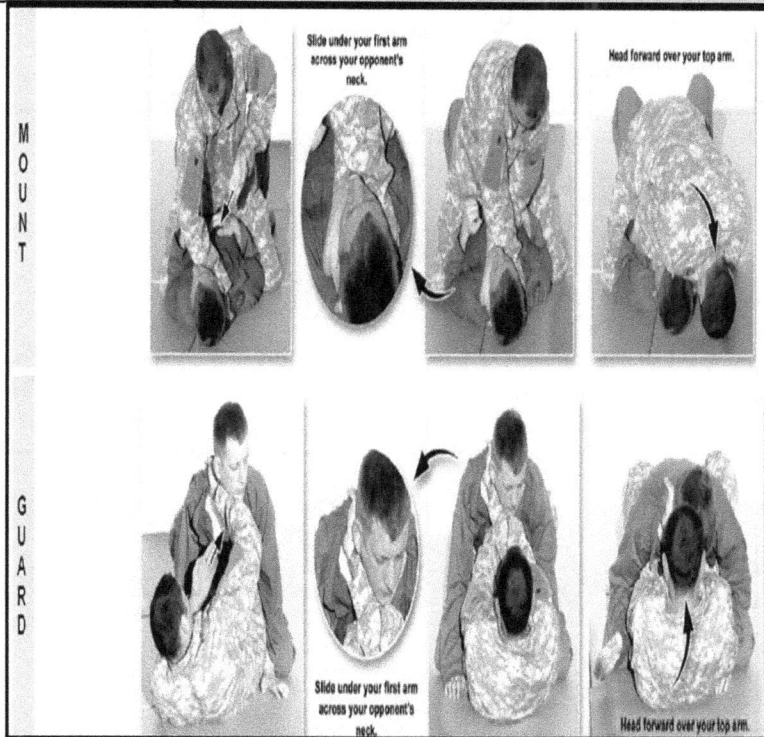

Figure 071-COM-0512-10. Cross Collar Choke Continued.

(1) Open your opponent's same-side collar With your non-dominant hand.

(2) Reach across your body, and insert your dominant hand into the collar you just opened.

(3) Relax the dominant hand, and reach all the way behind your opponent's neck.

(4) Grasp his collar with your fingers on the inside and your thumb on the outside.

(5) Release the grip of your non-dominant hand, and move your dominant-side forearm across your opponent's neck under the first arm, clearing his chin.

(6) Reach all the way back untilyour dominant hand meets the other hand using the same grip.

(7) Turn your wrists so that your palms face you, and pull your opponent into you.

(8) Expand your chest,and pinch your shoulders together.

(9) Bring your elbows to your hips to finish the choke.

c. Achieve the Bent Arm Bar from the Mount and Side Control. (Figure 071-COM-0512-11)

Note: The Bent Arm Bar is a joint lock that attacks the shoulder girdle. This technique can be employed from either the Mount or Side Control.

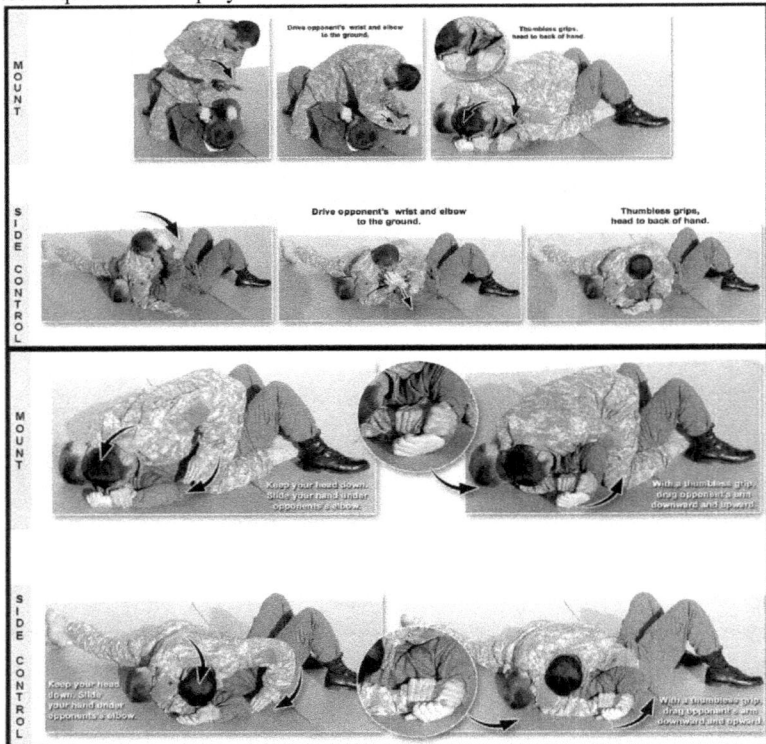

Figure 071-COM-0512-11. Bent Arm Bar.

(1) Drive your opponent's wrist and elbow to the ground with thumbless grip.

(2) Move your elbow to the notch created by your opponent's neck and shoulder.

(3) Keep your head on the back of your hand to protect your face from strikes.

(4) Place your other hand under his elbow.

(5) Grab your own wrist with a Thumbless Grip.

(6) Drag the back of your opponent's hand toward his waistline.

(7) Lift his elbow, and dislocate his shoulder.

d. Achieve the Straight Arm Bar from the Mount. (Figure 071-COM-0512-12)

Note: The Straight Arm Bar is a joint lock designed to damage the elbow. While this exercise outlines a Straight Arm Bar performed from the Mount, this technique can be performed from any dominant position.

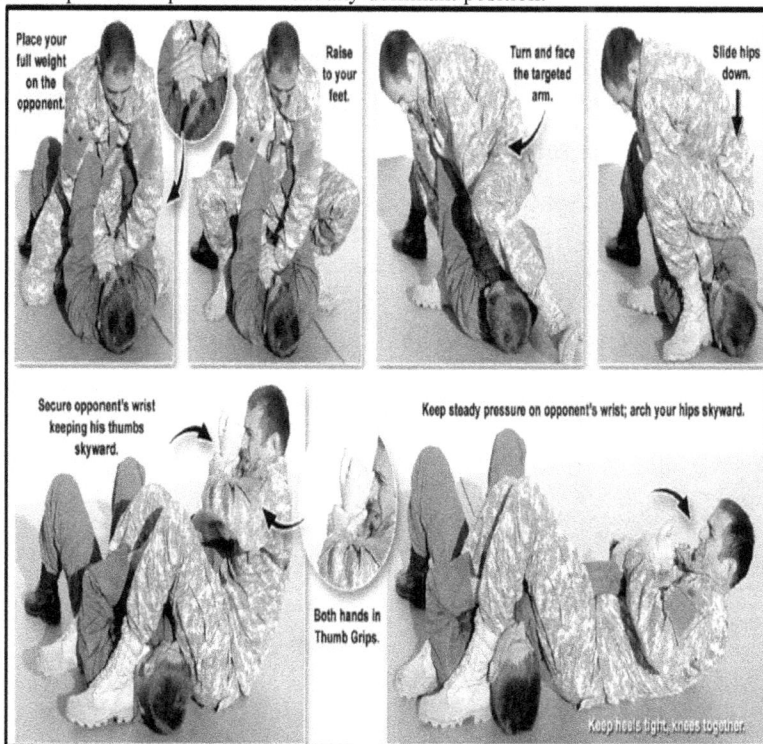

Figure 071-COM-0512-12. Straight Arm Bar from the Mount.

(1) Decide which arm you wish to attack.

(2) Isolate that arm by placing your opposite-side hand in the middle of your opponent's chest, between his arms.

(3) Target the unaffected arm and press down to prevent your opponent from getting off the flat of his back.

(4) Loop your same-side arm around the targeted arm and place that hand in the middle of your opponent's chest, applying greater pressure.

(5) Place all of your weight on your opponent's chest and raise to your feet in a very low squat.

(6) Turn your body 90 degrees to face the targeted arm.

(7) Bring the foot nearest to your opponent's head around his face, and plant it in the crook of his neck on the opposite side of the targeted arm.

(8) Slide your hips down the targeted arm, keeping your buttocks tight to your opponent's shoulder.

(9) Secure your opponent's wrist with both of your hands in Thumb Grips.

(10) Keep his thumb pointed skyward to achieve the correct angle.

(11) Pull your heels tight to your buttocks, and pinch your knees together tightly with the upper arm trapped between your knees, not resting on your groin.

(12) Apply steady pressure by trapping your opponent's wrist on your chest, and arching your hips skyward.

e. Achieve the Straight Arm Bar from the Guard. (Figure 071-COM-0512-13)

Note: Fighting from your back can be very dangerous. When your opponent attempts to strike and apply chokes from within your Guard, use the Straight Arm Bar from the Guard, a joint lock designed to damage the elbow.

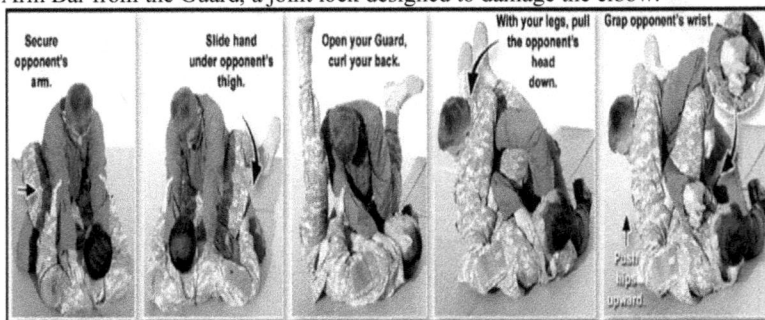

Figure 071-COM-0512-13. Straight Arm Bar from the Guard.

(1) Secure the arm at or above the elbow when your opponent presents a straight arm.

(2) Hold your opponent's elbow for the remainder of the move.

(3) Insert your other hand under the opponent's thigh on the side opposite the targeted arm.

Note: The hand should be palm up.

(4) Open your Guard, and bring your legs up, while curling your back to limit the friction.

(5) Contort your body by pulling with the hand that is on the back of your opponent's thigh.

(6) Bring your head to his knee.

(7) Place your leg over his head.

(8) Grab your opponent, and pull him down by pulling your heels to your buttocks and pinching your knees together with your leg.

(9) Move the hand that was behind your opponent's thigh to grasp the wrist that you secured at the elbow with a Thumb Grip.

(10) Curl your calf downward and push up with your hips to break your opponent's arm.

f. Achieve the Guillotine Choke. (Figure 071-COM-0512-14 and Figure 071-COM-0512-15)

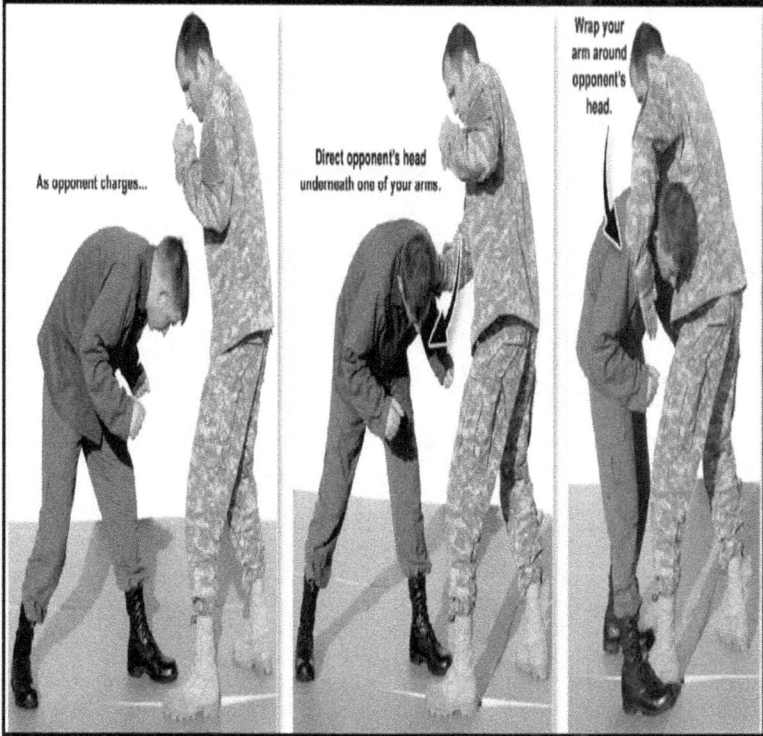

Figure 071-COM-0512-14. Guillotine Choke.

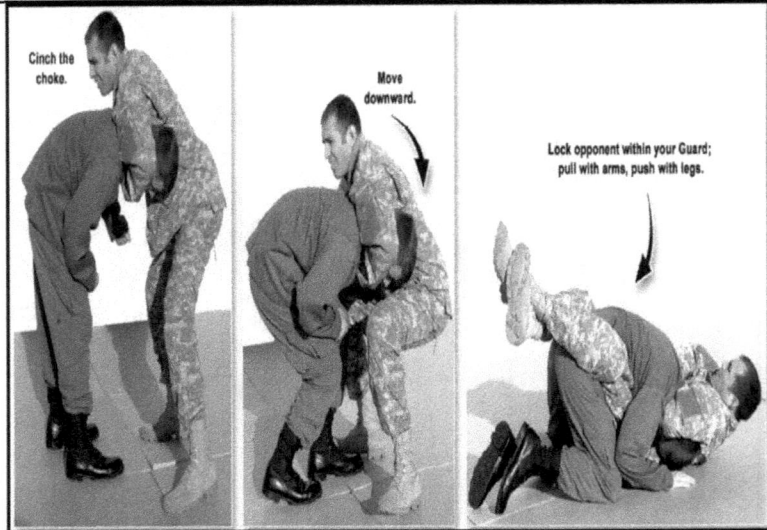

Figure 071-COM-0512-15. Guillotine Choke Continued.

    (1) Direct your opponent's head underneath one of your arms, and take a step back when your opponent charges your legs.

    (2) Wrap your arm around your opponent's head and under his neck.

    (3) Grasp the first hand where a watch would be with your other hand, ensuring that you have not reached around your opponent's arm.

    (4) Cinch the choke by bringing your arm further around your opponent's head, improving your grip.

    (5) Cinch up the choke and sit down to place him in your guard.

Note: Your palm should be facing your own chest.

    (6) Sit Down.

    (7) Place your opponent within your Guard.

    (8) Finish the choke by pulling with your arms and pushing with your legs.

Evaluation Preparation:

Setup: Provide the Soldier with the equipment and or materials described in the conditions statement.

Brief Soldier: Tell the Soldier what is expected of him by reviewing the task standards. Stress to the Soldier the importance of observing all cautions, warnings, and dangers to avoid injury to personnel and, if applicable, damage to equipment.

Performance Measures	GO	NO GO
1 Achieved the Clinch.	_____	_____
2 Gained a dominant position.	_____	_____
3 Finished the Fight.	_____	_____

Evaluation Guidance: Score the Soldier GO if all performance measures are passed. Score the Soldier NO-GO if any performance measure is failed. If the Soldier scores a NO-GO, show the Soldier what was done wrong and how to do it correctly.

References:
Required:
Related: TC 3-25.150

This page intentionally left blank.

APPENDIX A

Battle Drills

React to Contact:
Engage Targets with Individually assigned weapon
071-COM-0502 Move under Direct Fire
071-COM-0608 Use Visual Signaling Techniques
113-COM-1022 Perform Voice Communications
071-COM-0503 Move over, Through, or Around Obstacles (Except Minefields)
071-COM-0510 React to Indirect Fire while dismounted
071-COM-0513 Select Hasty Fighting Positions
071-COM-0501 Move as a member of a Fire Team
071-COM-4407 Employ Hand Grenades
071-COM-1271 Identify Visual Indicators of an Improvised Expolsive Device
071-COM-1270 React to Possible Improvised Explosive Devise

Establish Security at the Halt:
Engage Targets with Individually assigned weapon
071-COM-0801 Challenge Persons Entering your Area
071-COM-1004 Perform Duty as a Guard
071-COM-0815 Practice Noise and Light Discipline
113-COM-2070 Operate SINCGARS Single-Channel (SC)
113-COM-1022 Perform Voice Communications
171-COM-4079 Send a Situation Report (SITREP)
171-COM-4080 Send a Spot Report (SPOTREP)
071-COM-0513 Select Hasty Fighting Positions
071-COM-0608 Use Visual Signaling Techniques
052-COM-1361 Camouflage Yourself and Individual Equipment
071-COM-4408 Construct an Individual Fighting Position

Perform Tactical Combat Casualty Care:
081-COM-0101 Request Medical Evacuation
081-COM-1001 Evaluate a Casualty
081-COM-1005 Perform First Aid to Prevent or Control Shock
081-COM-1023 Open an Airway
081-COM-1046 Transport a Casualty
081-COM-1054 Evacuate Casualty
191-COM-0008 Search an Individual in a Tactical Environment
191-COM-0048 Apply a Combat Action Tourniquet (CAT)
081-COM-1055 Apply a Fox Eye Shield
081-COM-0013 Initiate a DD Form 1380 Tactical Combat Casualty Care (TCCC) Card
191-COM-1054 Apply an Emergency Bandage
191-COM-0099 Apply a Hemostatic Dressing
191-COM-0069 Apply a Occlusive Dressing

081-COM-1007 Perform First Aid for Burns
113-COM-1022 Perform Voice Communications

React to Ambush:

Near-

Engage Targets with Individually assigned weapon
052-COM-1271 Identify visual indicators of an IED
052-COM-3261 React to an IED attack
071-COM-0512 React to Hand-to-Hand Combat
071-COM-0030 Engage targets with M4/M16 Rifle
071-COM-4407 Employ hand grenades
071-COM-0501 Move as a member of a team
071-COM-0502 Move under direct fire
071-COM-0513 Select Hasty fighting positions
071-COM-0608 Use visual signal techniques
113-COM-1022 Perform voice communications

Far-

Engage Targets with Individually assigned weapon
052-COM-1270 React to an IED attack
071-COM-0501 Move as a member of a team
071-COM-0513 Select Hasty fighting positions
113-COM-1022 Perform voice communications
071-COM-0608 Use visual signal techniques
071-COM-0510 React to Indirect Fire dismounted
071-COM-0030 Engage Targets with individual weapon

Appendix B

Proponent School or Agency Codes

The first three digits of the task number identify the proponent school or agency responsible for the task. Record any comments or questions regarding the task summaries contained in this manual on a DA Form 2028 (*Recommended Changes to Publications and Blank Forms*) and send it to the proponent school with an information copy to:

Commander, U.S. Army Training Support Center
ATTN: ATIC-ITSC-CM
Fort Eustis, VA 23604-5166.

Table B-1. Proponent School or Agency Codes	
School Code	*Command*
MSCoE CM **031**	U.S. Army Chemical School Directorate of Training/Training Development 464 MANSCEN Loop, Suite 2617 Fort Leonard Wood, MO 65473-8929
MSCoE EN **052**	Commandant, U.S. Army Engineer School ATTN: ATSE-DT (Individual Training Division) 320 MANSCEN Loop, Suite 370 Fort Leonard Wood, MO 65473
FCoE **061**	Directorate of Training and Doctrine U.S. Army Field Artillery School ATTN: ATSF-D Fort Sill, OK 73503-5000
MCoE **071**	Commandant, U.S. Army Infantry School ATTN: ATSH-OTSS Fort Benning, GA 31905-5593
AHS **081**	Department of Training Support ATTN: MCCS-HTI 1750 Greeley Rd, Ste 135 Fort Sam Houston, TX 78234-5078
SCoE **091 (OMMS)** **093 (OMEMS)**	U.S. Army Combined Arms Support Command (CASCOM) Training Directorate USACASCOM, ATTN: ATCL-TD 2221 Adams Avenue., Suite 2018 Fort Lee, VA 23801-1809

Table B-1. Proponent School or Agency Codes	
School Code	*Command*
SCoE 101	Commander, US Army Quartermaster Center and School ATTN: ATSM-MA Fort Lee, VA 23801-5000
SCoE 113	Commander, USA Signal Center & School ATTN: ATZH-DTM-U Fort Gordon, GA 30905-5074
MCoE 171	Commander, USA Armor Center and School ATTN: ATZK-TDT-TD 204 1ST Cavalry Regiment Road Fort Knox, KY 40121-5123
JAG 181	Commandant, Judge Advocate General Legal Center and School ATTN: JAGS-TDD 600 Massie Road Charlottesville, VA 22903-1781
MSCoE MP 191	Commandant, United States Army Military Police School ATTN: ATSJ-Z 401 MANSCEN Loop, Suite 1068 Fort Leonard Wood, MO 65473-8926
APAC 224	Director, Army Public Affairs Center 6 ACR Road, Bldg 8607 ATTN: SAPA-PA Fort Meade, MD 20755-5650
ICoE 301	Commander, USA Intelligence Center & Fort Huachuca 550 Cibeque Street, Suite 168 ATTN: ATZS-TDS-I Fort Huachuca, AZ 85613-7002
JFK 331	U.S. Army JFK Special Warfare Center and School Fort Bragg, NC 28310-5000

Table B-1. Proponent School or Agency Codes	
School Code	*Command*
SCoE **551**	U.S. Army Combined Arms Support Command (CASCOM) Training Directorate USACASCOM, ATTN: ATCL-A 2221 Adams Avenue Fort Lee, VA 23801-2102
Fort Jackson **SRT**	U.S. Army Training Center, Fort Jackson Director of Basic Combat Training (DBCT), Doctrine and Training Development, ATTN: (ATZJ-DTD) 4325 Jackson Blvd. Fort Jackson, SC 29207-5315

This Page intentionally left blank.

GLOSSARY

Section I
Acronyms & Abbreviations

5-Cs	confirm, clear, call, cordon, and control
AAL	additional authorization list
ACE	air combat element (NATO);analysis and control element;armored combat earthmover;assistant corps engineer;aviation combat element (USMC); Avenger Control Electronics
ADA	air defense artillery; audio distribution amplifier; American Dietetic Association
AO	area of operations
APC	armored personnel carrier; activity processing code
AVPU	alertness, responsiveness to vocal stimuli, responsiveness to painful stimuli, unresponsiveness
BII	basic issue items
BSI	body substance isolation; Base Support Installation
CASEVAC	casualty evacuation
CBRN	chemical, biological, radiological, and nuclear
CPR	cardiopulmonary resuscitation
CSF	cerebrospinal fluid
CWIED	command wire improvised explosive device
DETCORD	detonator cord
EH	explosive hazards
EOF	escalation of force
EPW	enemy prisoner of war
F	frequency; fail; Fahrenheit; full; failed; Feeder; FMC
FM	field manual; frequency modulatedmodulation; flare multiunit; force module
FMC	full mission-capable; field medical card
FMI	field manual-interim; Failure Mode Identifier (indicates type of failure experienced by components. FMI has been adopted from SAE practice of J1587 diagnostics)
FOB	forward operating operationsoperational base; Free on Board
FW	Fixed Wing; Framework
GTA	graphic training aid
HQ	headquarters
IBA	individual ballistic armor
IED	imitative electromagnetic deception;improvised explosive device
JP	joint publication
LACE	liquid, ammunition, casualty, and equipment
LN; ln	local national; lane
MANPADS	man-portable air defense system

MEDEVAC	medical evacuation
METT-TC	A memory aid used in two contexts: (1) In the context of information management, the major subject categories into which relevant information is grouped for military operations: mission, enemy, terrain and weather, troops and support available, time available, civil considerations. (2) In the context of tactics, the major factors considered during mission analysis. [*Note:* the Marine Corps uses METT-T: mission, enemy, terrain and weather, troops and support available-time available.] (FM 6-0)
MIJI	meaconing, interference, jamming, and intrusion
MOI	message of interest; Material of Interest; memorandum of instruction; mechanism of injury
NBC	nuclear, biological, and chemical
NGO	nongovernmental organization; national government organization
NPA	net pay advice; nasopharyngeal airway
OAKOC	observation and fields of fire, avenues of approach, key terrain, obstacles, and cover and concealment
P	needs practice; pass; passed; barometric pressure; mean radius of curvature; positions; power; Propagated Booster; PMC
PBIED	person-borne improvised explosive device
PIR	priority intelligence requirements; priority information requirements
PZ	pickup zone
RCIED	radio controlled improvised explosive device
ROE	rules of engagement
RPG	rocket-propelled grenade
RTO	radio/telephone operator
RW	rotary wing; readwriter
SALUTE	size, activity, location, unit, time, and equipment
SMCT	Soldier's Manual of Common Tasks
SOI	signal operating/operation instructions
SOP	standing operating procedure
STP	shielded twisted pair; Soldier Training Publication; spanning-tree protocol; Soldier training plan
SURG	surgeon
SVBIED	suicide vehicle-borne improvised explosive device
TC	technical coordinator; training circular; track commander; tank commander; tactical commander; technical configuration
TCCC	tactical combat casualty care
TTP	tactics, techniques, and procedures
US	United States; ultrasound
VBIED	vehicle borne improvised explosive device
VOIED	victim-operated improvised explosive device

cont	continued; continuous; continuous fire; controlled substance
pnt	patient

This Page intentionally left blank.

REFERENCES

Required Publications

Required publications are sources that users must read in order to understand or to comply with this publication.

JOINT PUBLICATIONS
Most joint publications are available online at
www.dtic.mil/doctrine/new_pubs/jointpub.htm
DoD Dictionary of Military and Associated Terms, August 2017.
ACP 125 F. *Communications Instructions Radiotelephone Proceedures.* 5 September 2001.
http://www.navybmr.com/study%20material/ACP%20125.pdf
ACP 131. *Communications Instructions Operating Signals*, April 2009
http://wow.uscgaux.info/Uploads_wowII/114-01/Comms/ACP131F09_Operating_Z_O_Signals.pdf

ARMY PUBLICATIONS
Army regulations are available on the APD Web Site
(www.apd.army.mil)
ADP 3-0. *Operations.* 11 November 2016.
ADP 7-0. *Training Units and Developing Leader.* 23 August 2012.
ADRP 1-02. *Terms and Military Symbols*, 16 November 2016.
ATP 3-11.32/MCWP 3-37-.2/NTTP 3-11.37. *Multi-Service Tactics, Techniques, and Procedures for Chemical, Biological, Radiological, and Nuclear Passive Defense.* 5 May 2016.
ATP 3-11.37/MCWP 3-37.4/NTTP 3-11.29/AFTTP 3-2.44. *Multi-Service Tactics, Techniques, and Procedures for Chemical, Biological, Radiological, and Nuclear Reconnaissance and Surveillance.* 25 March 2013.
ATP 3-37.34. *Survivability Operations*, 28 June 2013
ATP 4-02.2. *Medical Evacuation.* 12 August 2014.
ATP 4-25.13. *Casualty Evacuation.* 15 February 2013.
ATP 5-19. *Risk Management.* 14 April 2014.
DA PAM 750-8. *The Army Maintenance Management System (TAMMS) User's Manual.* 22 August 2005.

ECBC-SP-036. *Guidelines for Mass Casualty Decontamination during an Hazmat/Weapon of Mass Destruction Incident: Volumes I and II.* August 2013. http://www.nfpa.org/~/media/files/news-and-research/resources/external-links/first-responders/decontamination/ecbc_guide_masscasualtydecontam_0813.pdf?la=en

FM 3-63. *Detainee Operations.* 28 April 2014.

FM 6-99. *US Army Report and message Formats.* 19 August 2013.

FM 21-60. *Visual Signal.* 30 September 1987.

Principles of Mass Casualty Decontamination. August 2013. *http://www.nfpa.org/~/media/files/news-and-research/resources/external-links/first-responders/decontamination/ecbc_guide_masscasualtydecontam_0813.pdf?la=en*

TB 9-2320-280-35-2 *Installation Instructions for Systems Single Channel Ground and Airborne Radio System (SINCGARS).* 17 September 2005.

TC 3-21.75. *The Warrior Ethos and Soldier Combat Skills.* 13 August 2013.

TC 3-22.6. *Guard Duty.* 13 January 2017.

TC 3-22.9. *Rifle and Carbine.* 13 May 2016.

TC 3-23.30. *Grenades and Pyrotechnic Signals.* 22 November 2013.

TC 4-02.1. *First Aid.* 5 August 2016.

TC 7-100. *Hybrid Threat.* 26 November 2010

TM 3-11.32. *Multi-Service Reference for Chemical, Biological, Radiological, and Nuclear Warning and Reporting and Hazard Prediction Procedures.* 15 May 2017.

TM 3-23.25. *Shoulder-Launched Munitions (This Item Is Published W/ Basic Incl. C1).* 15 September 2010.

TM 3-4230-229-10. *Operators Manual for Decontaminating Kit, Skin: M291, (NSN 4230-01-251-8702).* 2 October 1989.

TM 3-4230-235-10. *Operators Manual for Decontaminating Kit, Individual Equipment: M295 (NSN 6850-01-357-8456).* 21 November 2008.

TM 3-4240-542-13&P. *Operator and field maintenance Manual (including Repair Parts and Special Tools List) for Mask, Chemical-Biological: Joint Service General Purpose, Field, M50 Purpose, Field, M50.* 30 May 2008.

TM 3-6665-311-10. *Operators Manual for Paper Chemical Agent Detector: M9.* 31 August 1998.

TM 3-9905-002-10. *Technical Manual Operator's Manual for Sign Kit, Contamination: CBRN, M328 (NSN: 9905-01-567-7295) {NAVY SS200-A1-MMO-010}.* 10 November 2011

TM 9-1005-319-10. *Operator's Manual for Rifle, 5.56 MM, M16A2 (NSN 1005-01-128-9936), (EIC: 4GM) Rifle, 5.56 MM, M16A3 (NSN 1005-01-357-5112) Rifle, 5.56 MM, M16A4 (NSN 1005-01-383-2872) (EIC: 4F9) Carbine, 5.56 MM, M4 (NSN 1005-01-231-0973) (EIC: 4FJ) CARBINE, 5.56 MM, M4A1 (NSN 1005-01-382-0953) (EIC: 4GC).* 1 August 2016.

TM 9-1330-200-12. *Operators and organizational Maintenance Manual for Grenades {TM 1330-12/1A}* 17 September 1971.

TM 10-8415-209-10. *Operators Manual for Individual Chemical Protective Clothing.* 31 March 1993.

TM 10-8415-220-10. *Operators Manual for Joint Service Lightweight Integrated Suit technology (JSLIST) Chemical Protective Ensemble.* 28 July 2008.

TM 11-5820-890-13&P-1. *Operator and Field Maintenance Manual Including Repair Parts And Special Tools List For SINCGARS Ground ICOM Combat Net Radio: AN/PRC-119 (NSN: 5820-01-267-9482)(EIC: L2Q) AN/PRC-119D (NSN: 5820-01-421-0801)(EIC: GC9) AN/PRC-119F (NSN: 5820-01-451-8252)(EIC: GA4) AN/VRC-87A (NSN: 5820-01-267-9480)(EIC: L22) AN/VRC-87C (NSN: 5820-01-304-2045)(EIC: GDC) AN/VRC-87D (NSN: 5820-01-351-5259)(EIC: GAR) AN/VRC-87F (NSN: 5820-01-451-8248)(EIC: GA5) AN/VRC-88A (NSN: 5820-01-267-9481)(EIC: L23) AN/VRC-88D (NSN: 5820-01-352-1694)(EIC: GAS) AN/VRC-88F (NSN: 5820-01-452-8435)(EIC: GA3) AN/VRC-89A (NSN: 5820-01-267-9479)(EIC: L24) AN/VRC-89D (NSN: 5820-01-420-6619)(EIC: GD8) AN/VRC-89F (NSN: 5820-01-451-8247)(EIC: GAY) AN/PRC-90A (NSN: 5820-01-267-5105)(EIC: L25) AN/VRC-90D (NSN: 5820-01-420-6618)(EIC: GD9) AN/VRC-90F (NSN: 5820-01-451-8246)(EIC: GA2) AN/VRC-91A (NSN: 5820-01-267-9478)(EIC: L26) AN/VRC-91D (NSN: 5820-01-420-6621)(EIC: GDG) AN/VRC-91F (NSN: 5820-01-451-8249)(EIC: GA8) AN/VRC-92A (NSN: 5820-01-267-9477)(EIC: L27) AN/VRC-92D (NSN: 5820-01-421-2605)(EIC: GDH) AN/VRC-92F (NSN: 5820-01-451-8250)(EIC: GM2) (This item is included on EM 0239.* 1 August 2014.

TM 11-5820-890-13&P-2. *Operator and Field Maintenance Manual Including Repair Parts And Special Tools List For SINCGARS Ground ICOM Combat Net Radio:*

AN/PRC-119 (NSN: 5820-01-267-9482)(EIC: L2Q)
AN/PRC-119D (NSN: 5820-01-421-0801)(EIC: GC9)
AN/PRC-119F (NSN: 5820-01-451-8252)(EIC: GA4)
AN/VRC-87A (NSN: 5820-01-267-9480)(EIC: L22)
AN/VRC-87C (NSN: 5820-01-304-2045)(EIC: GDC)
AN/VRC-87D (NSN: 5820-01-351-5259)(EIC: GAR)
AN/VRC-87F (NSN: 5820-01-451-8248)(EIC: GA5)
AN/VRC-88A (NSN: 5820-01-267-9481)(EIC: L23)
AN/VRC-88D (NSN: 5820-01-352-1694)(EIC: GAS)
AN/VRC-88F (NSN: 5820-01-452-8435)(EIC: GA3)
AN/VRC-89A (NSN: 5820-01-267-9479)(EIC: L24)
AN/VRC-89D (NSN: 5820-01-420-6619)(EIC: GD8)
AN/VRC-89F (NSN: 5820-01-451-8247)(EIC: GAY)
AN/PRC-90A (NSN: 5820-01-267-5105)(EIC: L25)
AN/VRC-90D (NSN: 5820-01-420-6618)(EIC: GD9)
AN/VRC-90F (NSN: 5820-01-451-8246)(EIC: GA2)
AN/VRC-91A (NSN: 5820-01-267-9478)(EIC: L26)
AN/VRC-91D (NSN: 5820-01-420-6621)(EIC: GDG)
AN/VRC-91F (NSN: 5820-01-451-8249)(EIC: GA8)
AN/VRC-92A (NSN: 5820-01-267-9477)(EIC: L27)
AN/VRC-92D (NSN: 5820-01-421-2605)(EIC: GDH)
AN/VRC-92F (NSN: 5820-01-451-8250)(EIC: GM2)
(THIS ITEM IS INCLUDED ON EM 0239). 1 August
2014.

TM 11-5820-890-13&P-3. *Operator and Field Maintenance*
Manual Including Repair Parts And Special Tools List
For SINCGARS Ground ICOM Combat Net Radio:
AN/PRC-119 (NSN: 5820-01-267-9482)(EIC: L2Q)
AN/PRC-119D (NSN: 5820-01-421-0801)(EIC: GC9)
AN/PRC-119F (NSN: 5820-01-451-8252)(EIC: GA4)
AN/VRC-87A (NSN: 5820-01-267-9480)(EIC: L22)
AN/VRC-87C (NSN: 5820-01-304-2045)(EIC: GDC)
AN/VRC-87D (NSN: 5820-01-351-5259)(EIC: GAR)
AN/VRC-87F (NSN: 5820-01-451-8248)(EIC: GA5)
AN/VRC-88A (NSN: 5820-01-267-9481)(EIC: L23)
AN/VRC-88D (NSN: 5820-01-352-1694)(EIC: GAS)
AN/VRC-88F (NSN: 5820-01-452-8435)(EIC: GA3)
AN/VRC-89A (NSN: 5820-01-267-9479)(EIC: L24)
AN/VRC-89D (NSN: 5820-01-420-6619)(EIC: GD8)
AN/VRC-89F (NSN: 5820-01-451-8247)(EIC: GAY)
AN/PRC-90A (NSN: 5820-01-267-5105)(EIC: L25)
AN/VRC-90D (NSN: 5820-01-420-6618)(EIC: GD9)
AN/VRC-90F (NSN: 5820-01-451-8246)(EIC: GA2)
AN/VRC-91A (NSN: 5820-01-267-9478)(EIC: L26)
AN/VRC-91D (NSN: 5820-01-420-6621)(EIC: GDG)
AN/VRC-91F (NSN: 5820-01-451-8249)(EIC: GA8)
AN/VRC-92A (NSN: 5820-01-267-9477)(EIC: L27)
AN/VRC-92D (NSN: 5820-01-421-2605)(EIC: GDH)

AN/VRC-92F (NSN: 5820-01-451-8250)(EIC: GM2)
(THIS ITEM IS INCLUDED ON EM 0239). 1 August
2014.

TM 11-5820-890-13&P-4. *Operator and Field Maintenance
Manual Including Repair Parts And Special Tools List
For SINCGARS Ground ICOM Combat Net Radio:
AN/PRC-119 (NSN: 5820-01-267-9482)(EIC: L2Q)
AN/PRC-119D (NSN: 5820-01-421-0801)(EIC: GC9)
AN/PRC-119F (NSN: 5820-01-451-8252)(EIC: GA4)
AN/VRC-87A (NSN: 5820-01-267-9480)(EIC: L22)
AN/VRC-87C (NSN: 5820-01-304-2045)(EIC: GDC)
AN/VRC-87D (NSN: 5820-01-351-5259)(EIC: GAR)
AN/VRC-87F (NSN: 5820-01-451-8248)(EIC: GA5)
AN/VRC-88A (NSN: 5820-01-267-9481)(EIC: L23)
AN/VRC-88D (NSN: 5820-01-352-1694)(EIC: GAS)
AN/VRC-88F (NSN: 5820-01-452-8435)(EIC: GA3)
AN/VRC-89A (NSN: 5820-01-267-9479)(EIC: L24)
AN/VRC-89D (NSN: 5820-01-420-6619)(EIC: GD8)
AN/VRC-89F (NSN: 5820-01-451-8247)(EIC: GAY)
AN/PRC-90A (NSN: 5820-01-267-5105)(EIC: L25)
AN/VRC-90D (NSN: 5820-01-420-6618)(EIC: GD9)
AN/VRC-90F (NSN: 5820-01-451-8246)(EIC: GA2)
AN/VRC-91A (NSN: 5820-01-267-9478)(EIC: L26)
AN/VRC-91D (NSN: 5820-01-420-6621)(EIC: GDG)
AN/VRC-91F (NSN: 5820-01-451-8249)(EIC: GA8)
AN/VRC-92A (NSN: 5820-01-267-9477)(EIC: L27)
AN/VRC-92D (NSN: 5820-01-421-2605)(EIC: GDH)
AN/VRC-92F (NSN: 5820-01-451-8250)(EIC: GM2)
(THIS ITEM IS INCLUDED ON EM 0239).* 1 August
2014.

TM 11-5820-890-13&P-5. *Operator and Field Maintenance
Manual Including Repair Parts And Special Tools List
For SINCGARS Ground ICOM Combat Net Radio:
AN/PRC-119 (NSN: 5820-01-267-9482)(EIC: L2Q)
AN/PRC-119D (NSN: 5820-01-421-0801)(EIC: GC9)
AN/PRC-119F (NSN: 5820-01-451-8252)(EIC: GA4)
AN/VRC-87A (NSN: 5820-01-267-9480)(EIC: L22)
AN/VRC-87C (NSN: 5820-01-304-2045)(EIC: GDC)
AN/VRC-87D (NSN: 5820-01-351-5259)(EIC: GAR)
AN/VRC-87F (NSN: 5820-01-451-8248)(EIC: GA5)
AN/VRC-88A (NSN: 5820-01-267-9481)(EIC: L23)
AN/VRC-88D (NSN: 5820-01-352-1694)(EIC: GAS)
AN/VRC-88F (NSN: 5820-01-452-8435)(EIC: GA3)
AN/VRC-89A (NSN: 5820-01-267-9479)(EIC: L24)
AN/VRC-89D (NSN: 5820-01-420-6619)(EIC: GD8)
AN/VRC-89F (NSN: 5820-01-451-8247)(EIC: GAY)
AN/PRC-90A (NSN: 5820-01-267-5105)(EIC: L25)
AN/VRC-90D (NSN: 5820-01-420-6618)(EIC: GD9)*

AN/VRC-90F (NSN: 5820-01-451-8246)(EIC: GA2)
AN/VRC-91A (NSN: 5820-01-267-9478)(EIC: L26)
AN/VRC-91D (NSN: 5820-01-420-6621)(EIC: GDG)
AN/VRC-91F (NSN: 5820-01-451-8249)(EIC: GA8)
AN/VRC-92A (NSN: 5820-01-267-9477)(EIC: L27)
AN/VRC-92D (NSN: 5820-01-421-2605)(EIC: GDH)
AN/VRC-92F (NSN: 5820-01-451-8250)(EIC: GM2)
(THIS ITEM IS INCLUDED ON EM 0239). 1 August
2014.

TM 11-5820-890-13&P-6. *Operator and Field Maintenance*
Manual Including Repair Parts And Special Tools List
For SINCGARS Ground ICOM Combat Net Radio:
AN/PRC-119 (NSN: 5820-01-267-9482)(EIC: L2Q)
AN/PRC-119D (NSN: 5820-01-421-0801)(EIC: GC9)
AN/PRC-119F (NSN: 5820-01-451-8252)(EIC: GA4)
AN/VRC-87A (NSN: 5820-01-267-9480)(EIC: L22)
AN/VRC-87C (NSN: 5820-01-304-2045)(EIC: GDC)
AN/VRC-87D (NSN: 5820-01-351-5259)(EIC: GAR)
AN/VRC-87F (NSN: 5820-01-451-8248)(EIC: GA5)
AN/VRC-88A (NSN: 5820-01-267-9481)(EIC: L23)
AN/VRC-88D (NSN: 5820-01-352-1694)(EIC: GAS)
AN/VRC-88F (NSN: 5820-01-452-8435)(EIC: GA3)
AN/VRC-89A (NSN: 5820-01-267-9479)(EIC: L24)
AN/VRC-89D (NSN: 5820-01-420-6619)(EIC: GD8)
AN/VRC-89F (NSN: 5820-01-451-8247)(EIC: GAY)
AN/PRC-90A (NSN: 5820-01-267-5105)(EIC: L25)
AN/VRC-90D (NSN: 5820-01-420-6618)(EIC: GD9)
AN/VRC-90F (NSN: 5820-01-451-8246)(EIC: GA2)
AN/VRC-91A (NSN: 5820-01-267-9478)(EIC: L26)
AN/VRC-91D (NSN: 5820-01-420-6621)(EIC: GDG)
AN/VRC-91F (NSN: 5820-01-451-8249)(EIC: GA8)
AN/VRC-92A (NSN: 5820-01-267-9477)(EIC: L27)
AN/VRC-92D (NSN: 5820-01-421-2605)(EIC: GDH)
AN/VRC-92F (NSN: 5820-01-451-8250)(EIC: GM2)
(THIS ITEM IS INCLUDED ON EM 0239). 1 August
2014.

TM 11-5820-890-13&P-7. *Operator and Field Maintenance*
Manual Including Repair Parts And Special Tools List
For SINCGARS Ground ICOM Combat Net Radio:
AN/PRC-119 (NSN: 5820-01-267-9482)(EIC: L2Q)
AN/PRC-119D (NSN: 5820-01-421-0801)(EIC: GC9)
AN/PRC-119F (NSN: 5820-01-451-8252)(EIC: GA4)
AN/VRC-87A (NSN: 5820-01-267-9480)(EIC: L22)
AN/VRC-87C (NSN: 5820-01-304-2045)(EIC: GDC)
AN/VRC-87D (NSN: 5820-01-351-5259)(EIC: GAR)
AN/VRC-87F (NSN: 5820-01-451-8248)(EIC: GA5)
AN/VRC-88A (NSN: 5820-01-267-9481)(EIC: L23)
AN/VRC-88D (NSN: 5820-01-352-1694)(EIC: GAS)

AN/VRC-88F (NSN: 5820-01-452-8435)(EIC: GA3)
AN/VRC-89A (NSN: 5820-01-267-9479)(EIC: L24)
AN/VRC-89D (NSN: 5820-01-420-6619)(EIC: GD8)
AN/VRC-89F (NSN: 5820-01-451-8247)(EIC: GAY)
AN/PRC-90A (NSN: 5820-01-267-5105)(EIC: L25)
AN/VRC-90D (NSN: 5820-01-420-6618)(EIC: GD9)
AN/VRC-90F (NSN: 5820-01-451-8246)(EIC: GA2)
AN/VRC-91A (NSN: 5820-01-267-9478)(EIC: L26)
AN/VRC-91D (NSN: 5820-01-420-6621)(EIC: GDG)
AN/VRC-91F (NSN: 5820-01-451-8249)(EIC: GA8)
AN/VRC-92A (NSN: 5820-01-267-9477)(EIC: L27)
AN/VRC-92D (NSN: 5820-01-421-2605)(EIC: GDH)
AN/VRC-92F (NSN: 5820-01-451-8250)(EIC: GM2)
(THIS ITEM IS INCLUDED ON EM 0239). 1 August
2014.

TM 11-5820-890-13&P-8. *Operator and Field Maintenance
Manual Including Repair Parts And Special Tools List
For SINCGARS Ground ICOM Combat Net Radio:*
AN/PRC-119 (NSN: 5820-01-267-9482)(EIC: L2Q)
AN/PRC-119D (NSN: 5820-01-421-0801)(EIC: GC9)
AN/PRC-119F (NSN: 5820-01-451-8252)(EIC: GA4)
AN/VRC-87A (NSN: 5820-01-267-9480)(EIC: L22)
AN/VRC-87C (NSN: 5820-01-304-2045)(EIC: GDC)
AN/VRC-87D (NSN: 5820-01-351-5259)(EIC: GAR)
AN/VRC-87F (NSN: 5820-01-451-8248)(EIC: GA5)
AN/VRC-88A (NSN: 5820-01-267-9481)(EIC: L23)
AN/VRC-88D (NSN: 5820-01-352-1694)(EIC: GAS)
AN/VRC-88F (NSN: 5820-01-452-8435)(EIC: GA3)
AN/VRC-89A (NSN: 5820-01-267-9479)(EIC: L24)
AN/VRC-89D (NSN: 5820-01-420-6619)(EIC: GD8)
AN/VRC-89F (NSN: 5820-01-451-8247)(EIC: GAY)
AN/PRC-90A (NSN: 5820-01-267-5105)(EIC: L25)
AN/VRC-90D (NSN: 5820-01-420-6618)(EIC: GD9)
AN/VRC-90F (NSN: 5820-01-451-8246)(EIC: GA2)
AN/VRC-91A (NSN: 5820-01-267-9478)(EIC: L26)
AN/VRC-91D (NSN: 5820-01-420-6621)(EIC: GDG)
AN/VRC-91F (NSN: 5820-01-451-8249)(EIC: GA8)
AN/VRC-92A (NSN: 5820-01-267-9477)(EIC: L27)
AN/VRC-92D (NSN: 5820-01-421-2605)(EIC: GDH)
AN/VRC-92F (NSN: 5820-01-451-8250)(EIC: GM2)
(THIS ITEM IS INCLUDED ON EM 0239). 1 August
2014.

TM 11-5820-890-13&P-9. *Operator and Field Maintenance
Manual Including Repair Parts And Special Tools List
For SINCGARS Ground ICOM Combat Net Radio:*
AN/PRC-119 (NSN: 5820-01-267-9482)(EIC: L2Q)
AN/PRC-119D (NSN: 5820-01-421-0801)(EIC: GC9)
AN/PRC-119F (NSN: 5820-01-451-8252)(EIC: GA4)

AN/VRC-87A (NSN: 5820-01-267-9480)(EIC: L22)
AN/VRC-87C (NSN: 5820-01-304-2045)(EIC: GDC)
AN/VRC-87D (NSN: 5820-01-351-5259)(EIC: GAR)
AN/VRC-87F (NSN: 5820-01-451-8248)(EIC: GA5)
AN/VRC-88A (NSN: 5820-01-267-9481)(EIC: L23)
AN/VRC-88D (NSN: 5820-01-352-1694)(EIC: GAS)
AN/VRC-88F (NSN: 5820-01-452-8435)(EIC: GA3)
AN/VRC-89A (NSN: 5820-01-267-9479)(EIC: L24)
AN/VRC-89D (NSN: 5820-01-420-6619)(EIC: GD8)
AN/VRC-89F (NSN: 5820-01-451-8247)(EIC: GAY)
AN/PRC-90A (NSN: 5820-01-267-5105)(EIC: L25)
AN/VRC-90D (NSN: 5820-01-420-6618)(EIC: GD9)
AN/VRC-90F (NSN: 5820-01-451-8246)(EIC: GA2)
AN/VRC-91A (NSN: 5820-01-267-9478)(EIC: L26)
AN/VRC-91D (NSN: 5820-01-420-6621)(EIC: GDG)
AN/VRC-91F (NSN: 5820-01-451-8249)(EIC: GA8)
AN/VRC-92A (NSN: 5820-01-267-9477)(EIC: L27)
AN/VRC-92D (NSN: 5820-01-421-2605)(EIC: GDH)
AN/VRC-92F (NSN: 5820-01-451-8250)(EIC: GM2)
(THIS ITEM IS INCLUDED ON EM 0239). 1 August
2014.

TM 11-5820-890-13&P-10. *Operator and Field*
Maintenance Manual Including Repair Parts And
Special Tools List For SINCGARS Ground ICOM
Combat Net Radio: AN/PRC-119 (NSN: 5820-01-267-
9482)(EIC: L2Q) AN/PRC-119D (NSN: 5820-01-421-
0801)(EIC: GC9) AN/PRC-119F (NSN: 5820-01-451-
8252)(EIC: GA4) AN/VRC-87A (NSN: 5820-01-267-
9480)(EIC: L22) AN/VRC-87C (NSN: 5820-01-304-
2045)(EIC: GDC) AN/VRC-87D (NSN: 5820-01-351-
5259)(EIC: GAR) AN/VRC-87F (NSN: 5820-01-451-
8248)(EIC: GA5) AN/VRC-88A (NSN: 5820-01-267-
9481)(EIC: L23) AN/VRC-88D (NSN: 5820-01-352-
1694)(EIC: GAS) AN/VRC-88F (NSN: 5820-01-452-
8435)(EIC: GA3) AN/VRC-89A (NSN: 5820-01-267-
9479)(EIC: L24) AN/VRC-89D (NSN: 5820-01-420-
6619)(EIC: GD8) AN/VRC-89F (NSN: 5820-01-451-
8247)(EIC: GAY) AN/PRC-90A (NSN: 5820-01-267-
5105)(EIC: L25) AN/VRC-90D (NSN: 5820-01-420-
6618)(EIC: GD9) AN/VRC-90F (NSN: 5820-01-451-
8246)(EIC: GA2) AN/VRC-91A (NSN: 5820-01-267-
9478)(EIC: L26) AN/VRC-91D (NSN: 5820-01-420-
6621)(EIC: GDG) AN/VRC-91F (NSN: 5820-01-451-
8249)(EIC: GA8) AN/VRC-92A (NSN: 5820-01-267-
9477)(EIC: L27) AN/VRC-92D (NSN: 5820-01-421-
2605)(EIC: GDH) AN/VRC-92F (NSN: 5820-01-451-
8250)(EIC: GM2) (THIS ITEM IS INCLUDED ON EM
0239). 1 August 2014.

Related Publications

Related publications are sources of additional information. They are not required in order to understand this publication.

ARMY PUBLICATIONS

Army regulations are available on the APD Web Site (www.apd.army.mil)

AR 27-1. *Legal Services, Judge Advocate Legal Services.* 24 January 2017.

AR 40-66. *Medical Record Administration and Health Care Documentation.* 17 June 2008.

AR 40-400. *Patient Administration.* 8 July 2014.

ATP 3-22.40/MCWP 3-15.8/NTTP 3-07.3.2/AFTTP 3-2.45/CGTTP 3-93.2. *Multi-Service Tactics, Techniques, and Procedures for the Employment of Nonlethal Weapons.* 13 February 2015.

ATP 3-34.5. *Environmental Considerations.* 10 August 2015.

ATP 3-55.4. *Techniques for Information Collection during Operations among Populations.* 5 April 2016

ATP 5-19. *Risk Management.* 14 April 2014.

ATP 6-02.53. *Techniques for Tactical Radio Operations.* 7 January 2016.

ATTP 3-06.11. *Combined Arms Operations in Urban Terrain.* 10 June 2011.

DA Pamphet 750-8. *The Army Maintenance Management System (TAMMS) User's Manual.* 22 August 2005.

Department of Defense Law of War Manual (10 Soldier's Rules). *http://archive.defense.gov/pubs/Law-of-War-Manual-June-2015.pdf*

FM 2-22.3. *Human Intelligence Collector Operations.* 6 September 2006.

FM 27-10. *The Law of Land Warfare.* 18 July 1956.

Geneva and Hague Convention, *Laws of War.*
http://www.loc.gov/rr/frd/Military_Law/pdf/ASubjScd-27-1_1975.pdf

GTA 05-08-002. *Environmental-Related Risk Assessment.* 31 October 2013.

ISBN 9781284041750. *Prehospital Trauma Life Support (Military Edition).*
https://www.amazon.com/Prehospital-Trauma-Life-Support-Military/dp/1284041751.

STP 21-24-SMCT. *Soldier's Manual of Common Tasks (SMCT) Warrior Leader Skill Level 2, 3, And 4.* 9 September 2008.

TC 3-25.26. *Map Reading and Land Navigation.* 15 November 2013.

TC 3-25.150. *Combatives.* 31 March 2017.

TC 7-98-1. *Stability and Support Operations Training Support Package.* 5 June 1997.

Recommended Readings

Manual for Courts-Martial. 2016 Edition

> *http://jsc.defense.gov/Portals/99/Documents/MCM2016.*
> *pdf?ver=2016-12-08-181411-957*

Operational Law Handbook

> *https://www.loc.gov/rr/frd/Military_Law/operational-law-*
> *handbooks.html*

Prescribed Forms
None

Referenced Forms

Department of the Army Forms

DA Forms are available on the APD Web Site (www.apd.army.mil)

DA Form 2028. *Recommended Changes to Publications and Blank Forms.*

DA Form 2404. *Equipment Inspection and Maintenance Worksheet.*

DA Form 2823. *Sworn Statement.*

DA Form 4002. *Evidence/Property Tag.*

DA Form 4137. *Evidence/Property Custody Document.*

DA Form 5164-R. *Hands-on Evaluation LRA.*

DA Form 5165-R. *Field Expedient Squad Book.*

DA Form 5517. *Standard Range Card.*

DA Form 5988-E. *Equipment Inspection Maintenance Worksheet (EGA).*"Printed forms are available through normal supply channels.

Department of Defense Form

DD Forms are available on the OSD Web Site (http://www.esd.whs.mil/Directives/forms/)

DD Form 1380. *Tactical Combat Casualty Care (TCCC) Card.*

DD Form 2745. *Enemy Prisoner of War (EPW) Capture Tag.*

DD Form 2977. *Deliberate Risk Assessment Worksheet.*

By Order of the Secretary of the Army:

MARK A. MILLEY
General, United States Army
Chief of Staff

Official:

GERALD B. O'KEEFE
Administrative Assistant to the
Secretary of the Army
1727006

DISTRIBUTION:

Active Army, Army National Guard, and United States Army Reserve: To be distributed in accordance with the initial distrubution number (IDN) 111447, requirements for STP 21-1-SMCT.

www.ingramcontent.com/pod-product-compliance
Lightning Source LLC
Chambersburg PA
CBHW071318210326
41597CB00015B/1270